neuro

DR. A.K. PRADEEP · DR. ANIRUDH ACHARYA
DR. RAJAT CHAKRAVARTY · RATNAKAR DEV

neuroAI

Winning the Minds of Consumers
with Neuroscience-Powered GenAI

WILEY

Library of Congress Cataloging-in-Publication Data:

Names: Pradeep, A. K., 1963- author. | Chakravarty, Rajat, author. |
 Acharya, Anirudh, author. | Dev, Ratnakar, author.
Title: neuroAi : winning the minds of consumers with neuroscience powered
 GenAi / Dr. A. K. Pradeep, Dr. Rajat Chakravarty, Dr. Anirudh Acharya,
 Ratnakar Dev.
Description: Hoboken, New Jersey : Wiley, [2024] | Includes bibliographical
 references and index.
Identifiers: LCCN 2024020698 (print) | LCCN 2024020699 (ebook) | ISBN
 9781394261963 (hardback) | ISBN 9781394261987 (adobe pdf) | ISBN
 9781394261970 (epub)
Subjects: LCSH: Artificial intelligence—Marketing applications. |
 Consumers—Psychology.
Classification: LCC HF5415.125 .P73 2024 (print) | LCC HF5415.125 (ebook)
 | DDC 658.8/342028563—dc23/eng/20240601
LC record available at https://lccn.loc.gov/2024020698
LC ebook record available at https://lccn.loc.gov/2024020699

Cover Design: Wiley
Cover Image: © Africa Studio/Adobe Stock
SKY10079515_071724

Dr. A. K. Pradeep dedicates this book to his three adult kids – Alexis Pracar, Shane Pracar, and Devin Pracar – for keeping him curious, humble, and honest, to his Dad – Professor Anantha Krishnan – for his unwavering support, and to his Mom – Saraswathi as she sleeps on the waves on Stinson Beach, California. He thanks his sister, Dr. A. K. Bhargavi, and his brother, A. K. Prasad, for their unwavering support.

Dr. Anirudh Acharya dedicates this book to the wonderfully clever and dedicated colleagues he has had the pleasure of working with. Thank you for making every project an experience. This book would not have been possible without the love and patience of Reha.

Dr. Rajat Chakravarty dedicates this book to his wife, Dr. Torsa Ghosal – Here's to all the quirky moments that inspired these pages. Without you, this book would still be well-written but oh so dull.

Ratnakar Dev dedicates this book to Malhar and Arya, the future, and Sonali for all the support.

All authors together as a group dedicate this book to our dear friend Tom Robbins, for his infectious enthusiasm, joy, hard work, and sheer brilliance.

Contents

Introduction

There are moments in human history that are as rare as they are revolutionary. They change the very nature of life, largely for the good of all humankind.

While these historical moments are few, they create an outsized and lasting impact. Witness the printing press . . . the Industrial Revolution . . . the discovery of cures for diseases such as cancer, and several more.

We are witnessing just such a singular moment now: the explosive creation of generative Ai (GenAi). Never before has a technology swept across the globe so quickly, transforming virtually every aspect of life, from commerce to medicine to education, entertainment, governance, and even thought itself.

This chapter serves as a doorway through which you can pass, to encounter and understand the full sweep of this technology, and grasp its import and promise.

But this book introduces an even larger and more impactful event: the integration of GenAi with neuroscience. The implications of this unique combination are, in a very real sense, limitless.

Understanding how the human brain works will give you a new and insightful view into life and all its component aspects. Understanding GenAi, and even more specifically the interplay between neuroscience and AI, will take you well into the new world which we are all inhabiting more and more every day.

But this chapter, and this book, are much more than an elite lecture on generative Ai and neuroscience. The parallel purpose of this book, and each of its chapters, is to equip you with the knowledge and the tools that this revolutionary invention offers, that which you can use in your career and your life.

Given the proliferation of GenAi, it is important to ask the question, "if everyone has GenAi, how does it create a differentiator? Does generative Ai rapidly become generic Ai?"

We posit that the intelligent blending of neuroscience and GenAi creates true marketplace differentiators, and this the natural evolution of GenAi to neuroAi.

Welcome to the natural evolution of GenAi – neuroAi. It is natural to love the pretty pictures generated by GenAi, and to fall in love with the prose it generates, and falsely believe that humans are no longer necessary. This is just an arrogant, and erroneous premise, and this book refutes that. We believe that integrating the painstaking work of neuroscientists who have spent years deciphering the human brain is critical to the success of GenAi. We also assert that the painstaking work done by market researchers who have sought to understand the motivations of customers' needs to be humbly integrated into GenAi. The breakthrough work of cognitive scientists and psychologists in creating breakthroughs in understanding unconscious motivations cannot be brushed aside. Last, but not least, the thousands of professors, graduate students, and undergraduates who have submitted themselves to surveys, batteries of psychological assessments, EEG recordings, and fMRI analyses need to be acknowledged and form the guardrails and guiding principles of GenAi.

Humility in honoring and integrating these neuroscientific, psychological, and market insights findings into GenAi, does not water it down, but rather powers it to truly realize its full potential – this is the premise of this book.

Our approach to GenAi, furthermore, does NOT eliminate humans – it effectively carves a role for humans in an enterprise. We offer the following paradigm to those who seek to unleash the power of GenAi in an enterprise. If you walk in with the assumption that the single value of GenAi will be to bring your costs down by eliminating a chunk of your workforce – you will be proven wrong, and more so you will embark on a detrimental path.

This book is about top-line growth – how to create products that truly mesmerize and delight consumers. How to create services that blow their mind. How to inspire, challenge, and fulfill our deepest sensory cravings while growing revenues. The focus of this book is not on saving your way to prosperity by eliminating people and slimming costs through the use of neuroAi. We want you to be inspired to grow your revenues and win the mindshare of consumers. We therefore urge you to consider the following guidelines:

1. **Verify GenAi's understanding:** Use humans to verify if GenAi has understood what it has been asked to do perfectly. Humans migrate to verify and validate that the algorithm understands what it has been asked to do. In particular, confirm what its parameters, boundaries, and constraints are and how precisely it must deliver what is expected of it.

2. **Challenge and verify GenAi reasoning:** Use humans to verify and validate the reasoning GenAi has used to create its results – look for flaws and proof points that no constraints were violated and confirm that the output conforms to what was requested.

3. **Push GenAi creative domains:** Use humans to challenge and gently push the boundaries of GenAi and selectively explore newer areas that it was not quite programmed to explore. This is a place that humans excel at – extrapolation – and what better way to explore than with a machine that "knows it all."

4. **Transfer heuristics to GenAi:** Use humans to selectively transfer their "heuristics," "rules of thumb," and "life experience" as observations for GenAi. This preserves the expertise in an enterprise in a meaningful way, by embedding it into algorithms so it does not walk out the door when an employee leaves.

5. **Create institutional memory:** Use GenAi to create an enterprise memory of every single creative query, attempt, and conclusion reached by humans in the enterprise as the collective "creative treasure" for the next generation to use – so we never forget the lessons learned.

Our approach in this book is to give you the excitement of humanity embracing GenAi as a tool that honors our humanity, values it, and builds a better future for us with our understanding of ourselves.

Engaging Brains and Brands

Journey into the workings of the human mind and you'll find a labyrinth as complex as any plotline crafted by an avant-garde filmmaker. Our brains are wired for story, with each twist and turn of a narrative sparking connections across the convoluted neural networks. Now, imagine harnessing that power – not to perplex but to persuade. For marketers, this is not just a feat of creativity; it is a call to delve into the rich realm of neuroscience, because the nonconscious mind drives 95%+ of our decision-making (Kramer & Block 2011).

Here's where our tale takes a scientific spin: When we listen to stories, our brains light up not just in the language-processing regions, but across areas tasked with deciphering human emotions, motives, and experiences. This means the stories engulf us, resonating on a deeply personal level. An effective marketer sees not merely an opportunity but a responsibility to tell a tale that matters.

But how to hold that treasured attention? Contemporary evidence suggests that neural engagement peaks when humans can draw parallels between the story and their own lives, triggering a sense of personal relevance. As the science progresses, so does our understanding that the most compelling marketing narratives are those in which consumers can see themselves playing a starring role.

Transitioning these neuroscientific revelations into a practical marketing strategy invokes the delicate art of bridging dimensions: from firing synapses to firing up sales. The story must envelop the product in such a way that potential customers feel it was crafted just for them. Personalized marketing has shifted from being just a trend to becoming a necessity, justified by neuroscience (Montgomery & Smith 2009). It is important to teach GenAi that stories matter, and that is how humans perceive reality. Algorithms must learn how to create narratives that captivate attention, and drive resonance with emotion.

Employ subtle humor and watch as the guardrails of skepticism lower. Laughter, after all, is a universal language that the brain interprets as a signal of trust and camaraderie. When humor intertwines with a product's story, it can form an irresistible combination that charms the amygdala, the brain's bastion of emotion. Can algorithms really create humor? The breakthroughs in large language models (LLM) today make many things possible. Will they create the next Larry David or Dave Chappelle? That remains to be seen. But we urge you, dear reader, not to Curb your Enthusiasm.

Embark further down this neuroAi path, and we encounter the crucial element of sensory marketing. Research has shown the effectiveness of targeting multiple senses, cementing brand memories as undeniably as an unforgettable jingle or a haunting perfume. By engaging more senses, you stake a stronger claim in consumers' "neural real estate." GenAi algorithms need to understand that neurological resonance through multisensory stimuli are important for the brain. When authors begin a story with the sounds of a thunderstorm and smell of damp earth, they really engage neurological resonance.

In the world of experiences, where services and goods are staged like theatrical productions, we must consider the enveloping plot. Consumers are not mere observers but are part of the act. By curating immersive experiences, marketers are directors, orchestrating a play in which the product is a character, and the consumer the protagonist, deeply involved in the unfolding drama.

Let's not forget the power of visuals, which reign supreme in the kingdom of cognition. The human brain processes images quicker than words – an evolutionary trace to the days when survival hinged on instant recognition. In a market saturated with text, a well-crafted image resonates in a split second, striking the electric chords of our visual cortex. The brain has a complex set of rules developed over thousands of years on how best to parse visual imagery. The algorithms of GenAi must understand these rules to generate compelling and persuasive imagery, packaging designs, and point of sale materials.

Consider a product launch as a sort of first date. The initial impression matters greatly, and familiarity breeds affinity. Repetition in branding is not merely about consistency; it's about creating a rhythm that the brain grows fond of, building a comfort zone within which loyalty blossoms.

To recap as we conclude this narrative: Keep the tales woven, the humor clever, the senses captivated, and the visuals arresting. Make the mundane into the magical, for in the realm of neuroAi, the line between science and story is effectively blurred.

Driving Innovation by Mimicking Human Ingenuity

Human ingenuity is the masterful art of problem-solving, often breaking through established norms and practices. In neuroscience, our understanding of human behavior and the brain's mechanisms plays a pivotal role in advancing this art.

Neural mechanisms function as engines of cognition – intricate and profound, much like the awe-inspiring mechanics inside a high-performance automobile. And just as a sophisticated car demands a skilled driver, the complexities of the human brain require adept navigation. Current research into the frontal lobes, the "drivers" of executive functions, unveils the delicate interplay between cognition and control.

As we delve deeper into the neural realm, we recognize patterns and comprehend that the prefrontal cortex, akin to a dashboard laden with control buttons, regulates our thoughts, actions, and emotions with precision. The mastery lies not only in the impeccable design of this neural dashboard but also in the efficiency with which we use it to conduct our daily lives.

The concept of neuroplasticity presents the brain's remarkable ability to adapt – to learn how to drive itself more effectively over time, akin to refining one's driving skills continuously. This malleability is the cerebral counterpart to upgrades in vehicular technology, offering us the capacity to become better pilots of our mental faculties, constantly adjusting to the new routes of experiences and knowledge.

What really is innovation? Distilling and extracting lessons, ideas and breakthroughs from one field and creating their rich analogs in another field is indeed innovation. Scientific breakthroughs, product and service breakthroughs follow this paradigm. GenAi looks for breakthroughs in one area and thoughtfully asks what the analogs in another area are – this creates foundational breakthroughs in product innovation. It extracts breakthroughs and trends in one area and asks two fundamental questions:

1. Can these breakthroughs and trends be directly transferred into another area?
2. What are the analogs of this area that can be brought to the forefront into the other area?

These truly become the foundational paradigms of inspiration and innovation. The power of GenAi is the ability to perform this creative transfer across categories in an unceasing and unrelenting manner. Algorithms that create product innovation must take these evolutionary paradigms into creating product innovations across categories.

Protect the enterprise continuously by filing for patents and protections using the ability of GenAi to take a generated idea, and rapidly create a provisional patent. Imagine every brainstorming session coming up, not only with ideas for products, but a provisional patent accompanying the idea as well. The ability to create innovations, and protect them in real time is a powerful new use of GenAi.

Brains, Brands, and Bots

Most chatbots are boring, burdensome, and frustrating.

Most learning apps lose customers after two attempts or about two days of app usage.

Most healthcare and patient condition management applications are abandoned after just a few tries.

Billions of dollars are wasted in creating cold, unimaginative, unhelpful human machine interactions that rarely deliver on the promised benefits.

The burgeoning field of neuroscience offers an exquisite maze of knowledge, the understanding of which becomes a linchpin to building bridges between human cognition and artificial intelligence.

We begin with a dive into the unseen; the nonconscious processes that sculpt our interactions with the world. It's within this labyrinthine nonconscious that much of our decision-making takes center stage. Our neural processors handle a myriad of information, silently steering our likes, dislikes, and ultimately the choices we make, without the presence of conscious thought.

How do we get GenAi to understand how to seduce and persuade the nonconscious of the human mind? What data might it rely on to understand the nonconscious? These vexing puzzles have been solved well by generations of neuroscientists and marketers. We now know that music persuades the nonconscious mind, and so does reinforced entertainment – explaining reach, reinforcement, and effectiveness measures of content.

These nonconscious processes are pivotal when it comes to persuasion. Subtle cues, processed below the level of awareness, can have profound impacts on our choices. Understanding these subtle triggers, and the neural algorithms they set off, is key to creating messages that motivate and delight.

When it comes to coupling this intricate knowledge with the power of generative Ai, we open a portal of possibilities. AI, in its vast processing capabilities, can dissect, learn, and mimic the neural patterns that lead to these positive states of human experience. With AI as a partner, creative minds gain a powerful ally in their quest for innovation, engagement, and memorability.

The practical applications of this knowledge are manifold. Consumer experiences can be designed to be habit forming, to be delightful and truly exciting. Gone are the days of learning apps that fail and cold chatbots that bore. GenAi coupled with learning applications, and habit forming applications create newer possibilities for an informed healthier world. Passing the Turing test should not turn into passing the Boring test. Chatbots that engage, inspire, and motivate can create newer commercial opportunities.

Entertainment and sensory delights become tailored experiences that can adapt in real time to the neural feedback of the audience. Here, AI becomes an interactive companion, responding to the nonconscious cues of the participants, creating a symbiotic relationship between human and machine that elevates each performance to a new pinnacle.

The fusion of neuroscience and GenAi holds the keys to personalizing our world in ways we've only begun to understand. As this book unfolds, we merge deep scientific understanding with the spark of creativity and embark on a journey that reinvents how we think about the world of neuro-commerce.

The Artistry of AI-Augmented Creativity

One cannot help but marvel at the intricate interplay of neurons and synapses within the human brain, a duet that orchestrates our every thought, feeling, and creative burst. Our cerebral cortex permits us the luxury of abstract thought and ingenuity. With roughly 86 billion neurons, each functioning as a nexus of possibility, the potential for creative and intellectual achievements is boundless.

Indeed, the creative process is deeply rooted in the unpredictable yet harmonious interplay between various neural networks. The brain's default mode network (DMN), often active when we daydream or engage in introspection, plays a crucial role in generating novel ideas. It's within these moments of apparent rest that the seeds of creativity find fertile ground.

Transitioning into the world of augmentation, we find that generative Ai amplifies these capabilities. It adds a layer of complexity and depth. Machines learn and suggest, challenge, and inspire, functioning much like a muse to an artist, spurring human ingenuity to new heights.

As we harmonize our neural compositions with the algorithms of AI, the combination becomes a catalyst for creativity. Take the world of design. GenAi can sift through terabytes of visual art at superhuman speeds to inspire designers with themes, layouts, imagery and color palettes that might not surface in the isolation of the human mind. Algorithms can now create art "in the style of" that becomes a spectacular minimalistic extraction of what constitutes style. The ability to win with packaging in the aisle can change brand perception for both consumers and retailers.

Beyond suggestion, AI is also a co-creator. In music, algorithms analyze patterns in rhythm and melody, generating compositions that can evoke the deepest of human emotions – a blend of electronic and organic that enchants both creators and connoisseurs.

The beauty of AI collaboration lies in personalization. Marketing campaigns, once a one-size-fits-all suit, can now be tailored with precision only achievable by understanding and adapting to individual neural preferences. The result? A message that feels like it was crafted just for you, because, in a way, it was.

AI augments the creative process by introducing diversity and reducing the echo chamber effect. Content generation becomes less about rehashed ideas and more about unique narratives that resonate and connect. AI doesn't stifle creativity; it propels it, making laughter more heartfelt and surprises more delightful.

So, as we embark on this AI-assisted odyssey, let us remember that our neural narrators are not being silenced; they are being amplified. The generative prowess of AI serves as a powerful tool to refine and extend the narratives we wish to tell, painting our stories with a richer palette. AI is not the author of our tale; it is our sophisticated pen.

In essence, AI is a celebration, not a suppression, of our creative spirit. It stands as a testament to our ingenuity, a mirror that reflects our own cerebral prowess, stunningly amplified.

Building neuroAi Capability in the Enterprise Fast – Note for Leaders

The alliance between humans and machines is grounded in this principle. As we frame the context for our AI counterparts, the human brain finely hones the generative processes through strategic, creative prompts, guiding the technology akin to shaping clay on a pottery wheel, where each touch can alter the output. Thus, human involvement remains essential, ensuring the AI's extraordinary potential is channeled into products or solutions with meaningful applications.

When human touch converges with AI's capabilities, the outcome can often surpass what each could achieve in isolation. Consider the human-Doppler effect where, just as a siren's pitch varies as it passes by, the human contribution dynamically adjusts the direction and intensity of the AI's work based on real-time feedback and intuition.

Let's not forget that human creativity is not a static construct but a dynamic flow that often involves flashes of serendipity and leaps of imagination. It is these very facets that we entrust to our silicon partners.

Translating these neural principles into business strategizing, we find that human involvement with AI mirrors the core skills of any great CEO: anticipation, adaptation, and improvisation. CEOs often have to steer the company ship through turbulent waters – a knack equally required when calibrating AI to hit the sweet spot between ingenuity and practicality.

Imagine AI as a high-octane vehicle, the likes of which would leave any car lover filled with admiration. Now consider this: Just as you wouldn't hand your nephew the keys to a brand-new Ferrari on his sixteenth birthday, the reins to AI require a seasoned hand, someone who knows just when to throttle and when to brake.

AI without human input can generate content at breakneck speeds, but it's the human touch that adds depth, evoking the "oohs" and "aahs" from the audience.

The common approaches to building and enhancing GenAi capability in the enterprise are as follows:

1. Designate an executive as the GenAi leader within the enterprise – exec assembles multifunctional teams across disciplines.
2. Create a thousand points of light – by performing a variety of pilots across disciplines to figure where value is maximized within the enterprise.

3. Work with vendors of many sizes and capabilities to embed GenAi within the enterprise.

4. Hire some of the large, well-known consulting companies to lay out a GenAi strategy and implementation plan.

Reluctantly, we posit that most of these approaches will yield very small and mostly meaningless results with little or no ability to scale them. Sadly, millions of dollars will be spent with not much to show for it. The focus will be on unimaginative, brutal, and unnecessary layoffs and cost reductions that will not grow the top line, nor will it build for success.

However, there are two approaches that will work in superior and cost-effective ways, and will grow the top line, and we urge our reader, and business leader to consider them carefully.

1. **Acquire and build:** Rapidly acquire small GenAi or neuroAi companies that have domain knowledge, deep expertise, and proprietary data that range in size from 10 to 50 people. Integrate them into the enterprise and utilize their "entrepreneurial energy" and domain knowledge to transform the enterprise.

2. **Build-operate-transfer:** Use experienced vendors with deep knowledge of both the technology and the category to build neuroAi labs or Centers of Excellence, or neuroAi studios within an enterprise. Have them operate it with staff in the enterprise, and eventually transfer it to inheritors within the enterprise. Rotate employees through these in-house labs or studios to get the entire organization trained in the language and techniques of neuroAi. This will upscale the skill sets of the entire organization.

These two approaches will yield remarkable results, accelerating GenAi capability and knowledge within the enterprise, and will be most cost effective while delivering true business outcomes and value.

The Importance of the Nonconscious in GenAi

In the field of human cognition, the nonconscious mind plays the elusive but dominant partner – guiding, influencing, but seldom seen. Recent neuroscience posits that our nonconscious processes direct much of our cognitive output, from snap decisions to complex problem-solving tasks. Indeed, 95% of daily human decision-making is in the nonconscious.

The anatomical seat of the nonconscious lies within the vast networks of the brain, such as the basal ganglia, orchestrating movements and learning patterns without a whisper to our conscious awareness. The brain's nonconscious systems operate silently, efficiently processing vast amounts of sensory data, often solving puzzles we didn't consciously know were being attempted.

Think of the nonconscious as the über-efficient personal assistant who's always got your back, fielding calls you didn't know were coming. It's the nonconscious that ties your shoelaces while your conscious mind is still wrestling with which sock to put on first.

This nonconscious machinery works relentlessly and gracefully. Heuristics, the brain's innate shortcuts, allow it to function efficiently, guiding our daily navigation with minimal cognitive effort. Neuroscience indicates that these rapid-fire mechanisms are governed by neural circuits within subcortical structures, such as the amygdala and hippocampus, which process emotional and environmental stimuli long before our conscious mind has had a chance to catch up.

GenAi engines, thus enlightened with nonconscious data are equipped to deliver outputs that resonate with our own cognitive harmonies.

How to Use This Book

This book is for the practitioner – so do not read it from cover to cover. Read a chapter or two – think about it, reflect on it. See how the neuroscience principles make sense to you, in your own life. Look at things around you with this light of understanding.

Get on any one of the popular LLMs – large language model engines. Write a prompt or two asking it to produce some output without injecting neuroscience into it. Now inject the principles of neuroscience into the prompts giving the LLM precise instruction in what to do, and what to avoid. See for yourself how the output changes.

If you are a creative designer, creating products or innovations or packaging for a demography, read the neuroscience that governs the behavior of the demography in the first half of the book. Now go to the second half of the book and pull the relevant chapters or sections for the task at hand. Put it together and work with GenAi – you will see the difference.

We have deliberately kept the neuroscience application focused and accessible to the practitioner.

At the risk of being redundant, we have tried to make each chapter self-sufficient, so don't be surprised to see a few facts and concepts show up repeatedly. The repetition not only reinforces, but also makes each chapter a bit independent of the others – so you can pick up any chapter and be able to apply without the burden of all the prior knowledge.

In closing, the elegant interplay between the nonconscious mind and AI presents a frontier teeming with potential. By tracing the neural pathways that govern our cognition, we can sculpt AI systems that speak to the nonconscious mind of the consumer.

Why is having this knowledge so valuable now?

Because as AI, particularly GenAi, assumes more and more of a central role in marketing and product innovation, understanding how best to reach and persuade the nonconscious mind becomes vital and critical to winning the consumer.

Move on from here to learn even more. Remember clearly that humanity is at the center. Human understanding of ourselves must drive GenAi. Humans become supervisors, overseers, and drivers of this powerful engine. Humility, not hubris, will allow us to realize the real power of GenAi.

PART 1

Neuroscience and Ai

CHAPTER 1

neuroAi for Marketers, Product Designers, and Executives

In short order, LLMs will become one of the most current, important, and ongoing topics of conversation worldwide.

Why? And what are LLMs?

This chapter will take you through the inner workings of GenAi, including large language models – that is, LLMs. With this compelling technology rapidly encircling the globe and penetrating into every corner of commerce and personal life, it makes sense to attain a working grasp of the fundamentals. Soon, fluency with this knowledge will be as necessary, and expected, a basic job and life skill as is our daily reliance on digital communications.

So let's learn about transformers, encoders, decoders, tokens, and the ultimate goal post, the Holy Grail of AI: artificial general intelligence.

It used to be the running joke in AI – that a million monkeys pounding out random keystrokes would hardly produce Shakespeare. It appears that the primates in San Francisco have done just that. The question is *how*? This chapter lays out the core elements of transformer technology. The goal is to give you a rudimentary understanding of the elements that go into it, but also stimulate a desire in you to build your own transformer for your own industry – whether you make fragrance, flavors, music, or floral arrangements. Understanding transformers will facilitate a newer way to think of your enterprise data.

It has been the lament of CTOs and marketers that despite the numerous "data lakes" and "data warehouses" in an enterprise, nothing new has really come about that facilitated daily use and breakthrough discovery. Transformer tech, in conjunction with large language models, can provide that precise enterprise asset. We anticipate the next generation of tech consulting, and strategic consulting to be practices that house and structure data to build enterprise specific and enterprise proprietary transformer models.

Introduction to Large Language Models

Large language models (LLMs) have clearly marked a paradigm shift in the field of natural language processing (NLP) and artificial intelligence over the last couple of years. So much so that we seem to hear about a new acronymized LLM practically every other day, adding to the list of the now well-known models – GPTs (OpenAI, March 14, 2023; OpenAI, November 5, 2019; OpenAI, November 30, 2022), PaLM (Google AI, n.d.), LLaMA (Meta, Llama, n.d.), Gemini (Pichai S., Hassabis D., December 6, 2023), and the like. Their successful capture of so much of the world's attention has to do with how adept LLMs are at understanding, generating, and interacting with human language in ways that appear startlingly lifelike and human.

NLP models and chatbots are not new. In fact, the first chatbot can be traced all the way back to ELIZA, developed between 1964 and 1967 at MIT (Weizenbaum 1966), and even the GPT line of models can be traced backed to GPT-1 released back in 2018 (Radford et al. 2018). So, what has changed in the last couple of years? What magical threshold do we seem to have crossed? In this chapter we will introduce LLMs, and explore LLMs and the specialized niche they have carved out in the domain of human language understanding.

At the heart of the new breed of LLMs is a neural network architecture that was first introduced in a landmark paper "Attention is all you need" (Vaswani et al. 2017). This paper described the transformer model, which has been the key to revolutionizing the process of human language understanding. Before the transformer model, most NLP models, such as recurrent neural networks (RNNs) and convolutional neural networks (CNNs), relied heavily on the sequence of data. One word after another, and one sentence after another. This made it difficult for any of these models to efficiently capture long-distance dependencies between words or build an understanding on the semantic weight of words given their context.

To the novice reader, neural networks are simply yet another way of fitting models to input-output data. The same way, lines, curves, splines, polynomials, and regressors fit input data to output data, neural networks are just another way to perform the same input-output data mapping. The "branding" is that the structure of neural networks vaguely mimic and are inspired by how neurons fire in our brain. The notions of neurons, layers, firing potentials and thresholds, and weighted and reinforced connections between them is a mathematical model of our own biological computer – the brain. Simple neural networks have an input layer, an output layer, and a hidden layer in between. Deep learning networks have millions of layers and neurons. Training typically involves presenting matched input output pairs and letting the neural network learn to adjust its neuronal weights to create perfect mappings between inputs and outputs. The adjustment of the weights of neuronal connections is typically accomplished by choosing paths that minimize the predicted output error. So, with an error function, and training data, and a means to adjust the weights of neuronal connections (similar to adjusting the coefficients of a polynomial in curve fitting), the input output mapping is accomplished. A portion of the training dataset is generally reserved for the purpose of testing. The build neural network that has never seen the test data is then tested on that reserved dataset. If it predicts accurately the outputs given the inputs, then the network is declared as fully trained and the job is done. These of course were the early days. Nothing awesome came from these neural networks, until the invention of the transformer model.

The transformer model addressed these limitations by introducing the concept of "attention." We will go into this in some detail a little further on, but very briefly, the attention mechanism does not process words one by one but considers a weighted sum of all the words in a sentence simultaneously. These weights determine which words get the most of the model's "attention," and this allows the model to capture the relationships between words in the sentence.

In addition to this novel neural network architecture, the other factor that has fueled the LLM explosion is computational power and the vast quantities of data that these models churn through and digest in their training. The "large" in LLM isn't just for show after all. The size of an LLM model is typically measured in the number of parameters it has. These parameters are the various numerical values that the model can adjust to better improve its performance. In general, the more parameters, the more capable the model.

The evolution of LLMs has been characterized by a continuous expansion of these parameters, which has led to exponential improvements in performance. From models with millions to billions, and even trillions of parameters, the trajectory has been astonishing.

A Brief Chronology of LLM Evolution

1. 2018 – GPT (Generative Pretrained Transformer): Introduced by OpenAI. It has 117 million parameters (https://openai.com/index/language-unsupervised/)

2. 2018 – BERT (Bidirectional Encoder Representations from Transformers): Introduced by Google. It has 340 million parameters (Devlin *et al.*, 2019)

3. 2019 – GPT-2: Introduced by OpenAI, with 1.5 billion parameters (https://cdn.openai.com/better-language-models/language_models_are_unsupervised_multitask_learners.pdf)

4. 2019 – T5 (Text-to-Text Transfer Transformer): Introduced by Google in 2019. The "base" version has 220 million parameters and the "large" version extends up to 11 billion parameters (Roberts A., Raffel C., February 24, 2020)

5. 2020 – GPT-3: Introduced by OpenAI. It has a staggering 175 billion parameters (Li C., June 3, 2020)

6. 2022 – PaLM: Introduced by Google, with 540 billion parameters (https://research.google/blog/pathways-language-model-palm-scaling-to-540-billion-parameters-for-breakthrough-performance/)

7. 2023 – GPT-4: Introduced by OpenAI. A reported 1.76 Trillion parameters (Schreiner M., July 11, 2023)

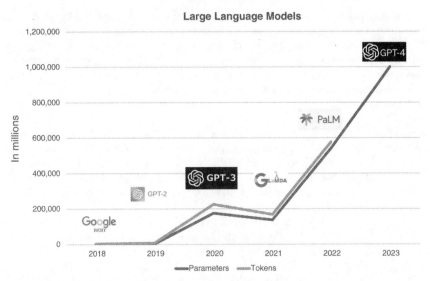

FIGURE 1.1 The rapid growth in size of recent LLMs.

Clearly, what is immediately obvious to anyone interacting with one of the newer LLMs is their ability to generate coherent and contextually relevant responses over extended interactions. This has made them particularly fascinating for both AI researchers and the general public, sparking conversations about the nature of intelligence, artificial general intelligence (AGI), and the future of human-AI interaction.

In the rest of this chapter, we will unpack the significance of LLMs, providing a walkthrough of the technical breakthroughs and what makes them special – especially the transformer models that underpin them and the accompanying operational mechanisms like tokenization and attention that allow for their unparalleled language understanding.

Transformers

As mentioned briefly, before the advent of transformers, the models in use were largely recurrent networks. The problem with RNNs, as we have seen, is that it tends to forget the beginning of the sentence by the time we come to the end of a very long sentence. Since these are sequential in nature, they process one word at a time. The output from the previous step is used as part of the input for the current step. This dependence on previous steps means that in an RNN each step must wait for the last one to be completed. This prevents parallelization during training and makes long sequences difficult to manage.

Transformers, on the other hand, have no recurrence and instead rely on *attention*, making it much faster to train and allows for parallelization. We will get into what the attention mechanism is shortly, but in general, such a mechanism allows a model to both process the entire input sentence simultaneously and "pay attention" to the important part of the sentence or input sequence.

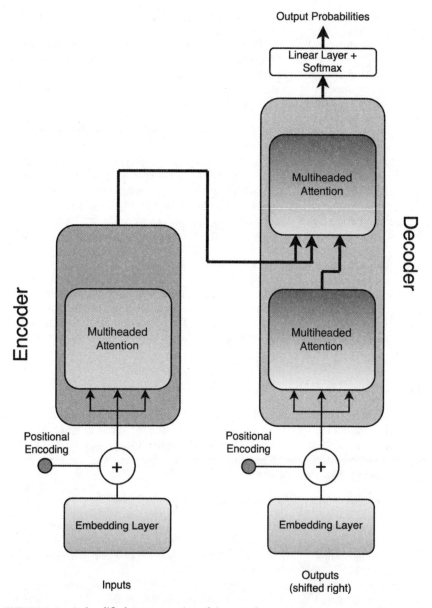

FIGURE 1.2   A simplified representation of the transformer architecture. Several intermediate layers have been hidden to focus on the key components.

Figure 1.2 depicts a transformer model. While it certainly looks to be complicated and has several components, at the highest level a complete transformer has two main blocks – an encoder block (on the left) and a decoder block (on the right). The encoder's job is to understand the input content, and the decoder's job is to utilize that understanding to produce the desired outcome, whether this be text generation, machine translation, question answering, or what have you.

An encoder can be thought of as the part of the model that changes or transforms the input and encodes it into a higher, more abstract mathematical representation that the model can digest. The purpose of the encoder is to extract and encode important features from the input that will be useful for the task at hand. In the case of text data, the encoder would take a sentence and convert each word into a mathematical entity called a vector that captures its semantic meaning and contextual relationship with other words in the sentence.

The output of the encoder is then passed to the decoder. The decoder takes the encoded input and generates a readable or useful output from the encoded information. For instance, in a translation task, the encoder will take in sentences of the source language and generate an abstract internal representation capturing the meaning and nuances of the input sentence. The decoder will then take this internal representation and generate a sentence in the target language, word by word. It uses the encoded information and also takes into account what has been translated so far, to generate the next part of the output.

A closer look at the transformer architecture reveals several subblocks marked with the words *attention* and *multiheaded attention*. We shall now take a closer look at the attention mechanism that forms the heart of the transformer architecture and explain how it knows what the important part of the sentence is and what it means for a model to pay attention to parts of a sentence.

The Self-Attention Mechanism

The self-attention mechanism in language models, in its most intuitive sense, allows the model to "focus" on the relevant parts of the input to make accurate predictions. It's akin to how when we read a complex text, we pay more attention to certain keywords to comprehend its meaning.

Let us consider the sentence:

"The quick brown fox jumps over the lazy dog."

When you parse through this sentence and notice the word "fox," you are also able to identify the words that would be most related to the words fox. In this case, they might be "quick," "brown," and "jumps." These other words help you understand what this particular fox is about.

So if you were to construct a **query**, *"What is the fox in this sentence about?"* the **keys** that would help you unlock the puzzle would be the words "quick," "brown," and "jumps." And each of these words carry its own **values** and associations. You have an understanding of what it means to be "quick" or to "jump," and therefore an understanding of what this fox is about.

Similarly, what the multiheaded attention mechanism does is to subject each word of the input sentence to the following sort of questioning process:

*"For the word **q** in the input sentence, what are the other words **k** in the sentence that best help me understand it, and what are their associated values **v**?"*

Through extensive training, the attention mechanism learns how to effectively and correctly represent **q**s, **k**s, and **v**s for itself, so that in any given sentence for each of the **q**s, it pays attention to the correct **k**s and retrieves the correct **v**s.

Let us build on this intuition and make it slightly more concrete. The next section is slightly more complex, and there will be some math involved.

The Details: Tokenization and Embeddings

The first thing we need to sort out before we proceed is to understand how a transformer model represents words mathematically. Let us consider the same sentence:

"The quick brown fox jumps over the lazy dog."

The typical procedure is that this sentence is first "tokenized" into a sequence of tokens or entities. A word-level tokenization would give us the following sequence:

"The", "quick", "brown", "fox", "jumps", "over", "the", "lazy", "dog"

Tokens   **Characters**

9   43

The quick brown fox jumps over the lazy dog

TEXT   TOKEN IDS

These tokens are all independent entities; across an entire corpus of data there would be a large, but limited, set of tokens that together define a *vocabulary*. This would be like a special dictionary where each word would be assigned a specific and unique token number (OpenAI "Tokenizer" n.d).

791, 4062, 14198, 39935, 35308, 927, 279, 16053, 5679

Tokens Characters

9   43

```
[791, 4062, 14198, 39935, 35308, 927, 279, 16053, 5679]
```

TEXT **TOKEN IDS**

Each token number would map to a high-dimensional mathematical vector called the *embedding vector* or *word vector*. The importance of embedding vectors is that via training, words/tokens are converted into vectors in a way that semantic relationships between words relate to the geometric relationships between their corresponding vectors. For instance, words with similar meanings are located near each other in the vector space, and the geometric direction between words can even capture semantic relationships between words. Remember the neuroscience adage: "Neurons that fire together, wire together." Words that seem connected seem to cluster together in the vector space of words.

The process of converting discrete words into vectors is called *embedding* or *word embedding*.

In a useful embedding, the closer two vectors are, the more related they are semantically. This measure of closeness is mathematically expressed as the angle between two vectors measured via the dot product. The smaller the angle, the more the vectors point in the same direction, the more they are semantically related. For vectors with unit length, we see their dot product is

$$\left(\mathbf{e,f}\right) = \cos\left(\mathbf{theta}\right),$$

where theta is the angle between them. The value of the dot product ranges from 1 to –1; an angle of 0 gives us a $(\mathbf{e,f}) = \mathbf{1}$, an angle of 90 degrees gives $(\mathbf{e,f}) = \mathbf{0}$, and an angle of 180 degrees gives –1.

This is how the machine mathematically represents the relationships between words – by constructing such an embedding space through training.

We have seen how the words are tokenized and converted into mathematical objects called embedding vectors. There is still a tiny piece missing.

Since the entire sentence is fed concurrently to the encoder, unlike the case of recurrent networks, the transformer model has no intrinsic understanding of the position of each word in the sentence. Therefore, position information must be injected into the input in some manner. This is done using positional encoding. In essence, a position-dependent vector is added to each of the input embedding vectors. We won't go into the details of how this positional encoding is constructed here, but for our purposes, it is sufficient to note that positional information is cleverly added to the input embedding vectors before it is fed into the encoder (Vaswani et al. 2017).

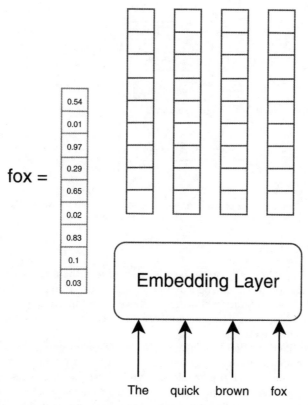

FIGURE 1.3   A representation of how the embedding layer converts words and their tokens into continuous valued vectors.

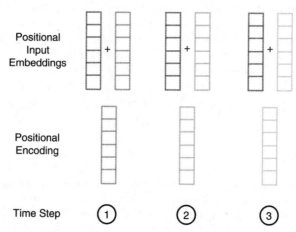

FIGURE 1.4   The addition of positional information to the input vector embeddings.

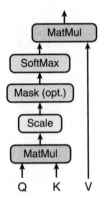

FIGURE 1.5 The internals of the attention mechanism. From Vaswani et al. 2017.

The Encoder

This set of embeddings (word + position) is fed into the encoder block. Let us recall that the job of the encoder is to map this input into an abstract internal mathematical representation that holds the semantic information for the input sequence. The first thing we encounter within the encoder block is the multi-headed attention layer.

The multiheaded attention layer applies the self-attention mechanism several times independently to the input sequence. As we have noted, the self-attention mechanism is the heart of the transformer and allows the model to selectively pay attention to the most important parts of the input sequence and to understand the meaning of a word within the context of the entire input sequence. Let us sketch out the mathematical operations underlying this.

The first thing that is done before the self-attention mechanism is to convert each of the input embedding vectors into three other vectors – the query (**q**), key (**k**), and value (**v**) vectors. This conversion occurs by passing the input embeddings through three separate and independent linear layers. A linear layer is a basic type of neural network connection, and the process of passing a vector through this layer is mathematically equivalent to multiplying each of the vectors with a matrix. These matrices are typically denoted **Wq**, **Wk**, **Wv**. If you are so inclined, this operation would mathematically be denoted by the equations:

$$q\_i = e\_i * (\mathbf{Wq})$$
$$k\_i = e\_i * (\mathbf{Wk})$$
$$v\_i = e\_i * (\mathbf{Wv})$$

The multiplication of each embedding vector **e_i** with the matrices **Wq**, **Wk**, **Wv** represents a linear transformation of the embedding vector into three different representations, **q_i**, **k_i**, and **v_i**. In a well-trained transformer, these linear transformations are carefully tuned so that each of these new vector representations correspond in a meaningful way to the semantic relationships between different words.

All the values of these matrices **Wq**, **Wk**, **Wv** will be learned by the model during training. At initialization, they are randomly assigned random values.

Let us pause now and describe what each of these query, key, and value vectors are, and how they play a role in the self-attention mechanism. Please keep the intuitive explanation provided earlier in mind as you read through the rest of this:

1. **Query:** A query vector **q** is a transformation of **e** by which the attention mechanism will try to find the most relevant tokens in the input sequence it should give more weight to. We could imagine this as a word in a sentence looking around to see which other words it should pay attention to.

2. **Key:** A key vector **k** is a different transformation of **e** that will be used by the attention mechanism to compute a match with the query vector **q**. A given query **q** is compared to all keys **k**, and the ones with the greatest match tell the transformer to weigh the corresponding tokens the most.

3. **Value:** A value vector **v** is a third transformation of **e**, into a space where relationships between the vectors **v** correspond to the semantic relationships between the tokens.

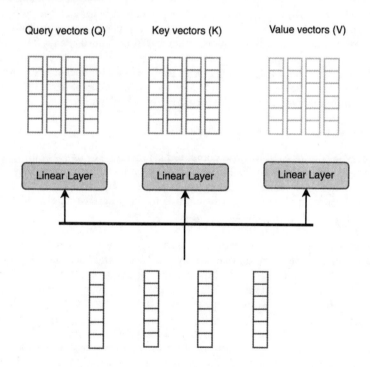

FIGURE 1.6   Linear transformation of embeddings into query, key, and value vectors.

	The	quick	brown	fox		dog
The	0.3	0.1	0.2	0.35		0.02
quick	0.03	0.4	0.1	0.23		0.01
brown	0.022	0.034	0.6	0.32	...	0
fox	0.1	0.15	0.2	0.5		0.05
dog	0.01	0.005	0.002	0.1		0.7

FIGURE 1.7   An illustrative attention matrix. The values represent the attention weights that inform the model of the context of each token.

The reason there are three separate transformations of the embedding **e** can be understood with the following example. Let us consider the query vector **q** for the word "apple." We would clearly like this vector to match well with both the key vectors **k** for the words "phone" and "fruit." This would mean that the two key vectors for "phone" and "fruit" would also have to match each other reasonably well. However, it is clear that the actual values of the two words should be quite dissimilar in general. Meaning the value vectors **v** should have a very low match. Using the same representation for queries, keys, and values, this would not be possible.

Each query vector **q** is matched with every key vector **k** via the dot product (**q**,**k**) to arrive at a set of numbers that determines the level of the match. A higher number implies a better match. So, for example, in our sentence

"The quick brown fox jumps over the lazy dog."

the query vector for each token is compared against the key of every token in the sentence. This would result in a matrix of dot product values, with queries along the rows, and keys along the columns. In our example, when the transformer is properly trained, the query corresponding to the token fox would have a high match with "quick," "brown," and "jumps."

This operation is the "MatMul" block in the attention diagram. The values in this matrix are scaled in such a way that the values in every row sum to 1. This is represented in the attention diagram by the "scale" and "SoftMax" layers. This matrix contains the attention weights for the particular input sequence, telling the model what to pay attention to for each part of the input sequence, knowing that to understand the fox here, one needs to pay attention to "quick," "brown," and "jumps."

In the next stage, this attention weight matrix is multiplied with the value vectors **v** from earlier. The multiplication of a value vector **v** by this weight matrix has the geometric effect of shifting its direction. Applied to all the value vectors, this operation results in a new set of output vectors for each of the

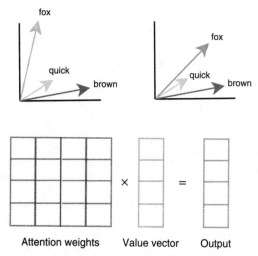

Attention weights   Value vector   Output

FIGURE 1.8   The output vectors from the attention mechanism are obtained by multiplying the value vectors V with the attention weight matrix.

tokens that are all directionally realigned. These new output vectors are the model's semantically weighted internal representation of each of the elements of the input sequence.

An intuitive way to think of this would be that the original input vector for "fox" $\mathbf{qf}$, is now, through the process of self-attention, directionally shifted to a new vector $\mathbf{q'f}$ that is a mathematical representation of "quick brown jumping fox", because the model has paid attention to and weighted the tokens "quick," "brown," and "jumping."

This is done for all the tokens in the input sequence, meaning that one could imagine the input vector for "dog" being shifted toward a new vector that in the internal mathematical space of the transformer stands for "lazy dog."

The transformer model learns the values of all of these embedding vectors and attention matrices through an extensive training process and feedback mechanism.

This entire sequence of operations just described constitutes a single "self-attention head." The multipart of the multiheaded attention indicated that this self-attention procedure is applied several independent times by repeated self-attention heads. These attention heads are all kept independent so that each can choose to learn a different representation of the input sequence. A loose analogy would be to say that multiple attention heads allow for the model to consider multiple interpretations of an input sequence.

In summary, the encoder block of the transformer uses multiheaded attention to compute attention weights for the input and produces a set of output vectors that encodes information about how each token relates to all the other tokens in the input sequence. This set of output vectors can be thought of as an internal mathematical representation of the contextual and semantic content of the input sequence.

The Decoder

The decoder of a transformer model plays a pivotal role in transforming the encoded input, specifically key and value vectors, into the desired output sequence. At a high-level view, the essence of the decoder revolves around being a predictive structure, operating an autoregressive generation of output tokens one by one based on the previously predicted tokens. We won't go into the details of the decoder architecture, but just note that it contains a couple of layers implementing self-attention, with a few feed-forward and linear layers at the end.

The first step in the decoding process involves feeding the decoder a special token called the start of sentence token **<sos>** that tells the decoder to start generating an output sequence. This **<sos>** token is embedded in the same manner as the input sequence was before being fed to the encoder.

We have seen that the key-value pairs that the encoder transmits are expressions of similarity relationships between input words, encapsulating the context from the input sequence. The decoder takes the query vector for the generated tokens (at the beginning, this would only be the single **<sos>** token), and the attention mechanism measures the relationships between the queries (the generated tokens) and all the key-value pairs from the input sequence fed to it by the encoder.

This quantification is used to weigh the values from the input sequence, ultimately creating a set of context vectors that fundamentally signifies the context that the model needs to focus on when generating the next output token.

The final linear layer in the decoder acts as a classifier. It takes the set of context vectors from the decoder and converts it into a probability distribution over all the tokens in the vocabulary. A token with the highest probabilitly can be picked to be the next predicted output in the sequence.

Once the first output token is generated (after the **<sos>** token), it becomes part of the input to generate the next token. This is the phase of autoregressive prediction. Here, the process works in a loop, with already predicted tokens influencing the prediction of the next token. By autoregressively leveraging the generated context vectors and previous output tokens, the decoder can successively produce the whole output sequence, one token at a time, until it generates an end token.

Thus, the decoder, equipped with a powerful loop of self-attention and context-aware prediction, consumes the encoded input and uses it to emit an output sequence with context awareness, precision, and consistency. This output sequence can be a next-word prediction to continue a sequence, or even a sequence-to-sequence translation task from one language to another.

Training the LLMs

The training of a transformer model, just like its architecture, isn't a straightforward process. It requires careful selection of data, precise application of masking and encoding techniques, patience, and a lot of computing power to expedite the process. In this chapter, we will dive deep into how a transformer is trained, why certain methodologies are adopted, and how this meticulous process captures and enhances the transformer's unparalleled capabilities.

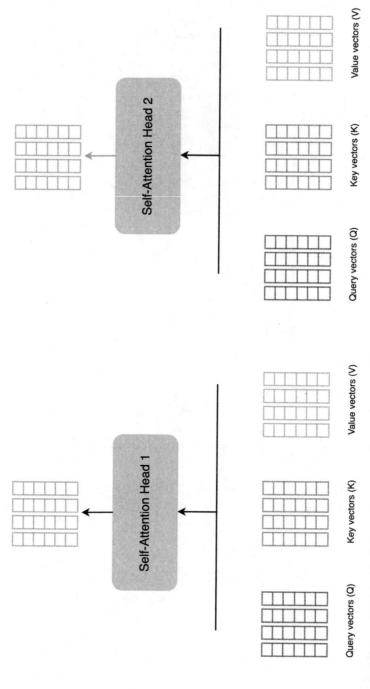

FIGURE 1.9 Multiheaded attention mechanism is a stacking of several single attention heads.

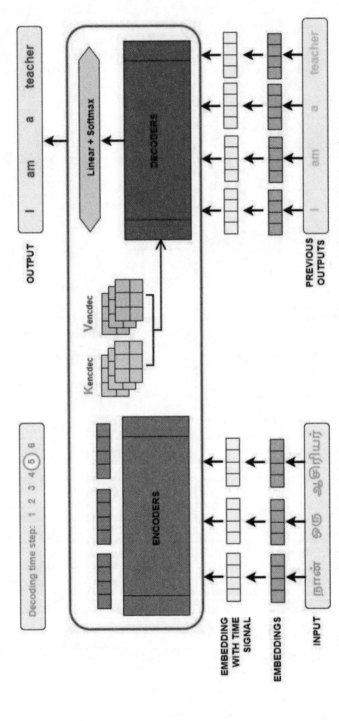

FIGURE 1.10 The decoder autoregressively generates the output in a loop, where the output of time step t is dependent on the previously generated tokens at time steps $1 \ldots t - 1$.

Data Collection

All machine learning models thrive on data, and transformer models are no exception. The type of data used for training is crucial, dictating the model's performance and utility. For instance, training GPT-3 utilized an enormous trove of data drawn from the internet at large, incorporating both language and programming code. The idea behind this was to develop a model that has a comprehensive understanding of human languages as well as programming languages. The linguistic data included a variety of text sources such as the book corpus, common crawl, Wikipedia, articles, and websites – a dataset comprising nearly a trillion words (Ali F., April 11, 2023). This enabled GPT-3 to grasp the nuances, dialects, and subtleties across different forms of language. Alongside this, the model was also trained on substantial amounts of programming code. This exposure allowed GPT-3 to learn the syntax, semantics, and logic of different programming languages, empowering it to generate code snippets when prompted.

Transformer Training

Training a transformer revolves around its fundamental components, the encoder and decoder. Both use self-attention mechanisms that parallelize the training process and facilitate the model's understanding of language elements. Optimizers and loss functions are employed during training, guiding the model toward better predictions by minimizing the difference between the predicted and actual outcomes.

The training procedure also involves special practices like masked encoder training and encoder-decoder paired training. In masked encoder training, some of the input data is intentionally masked or hidden during training. This practice prompts the model to predict the masked data based on the context, enhancing its proficiency in understanding and generating language.

Masked token training, often referred to as masked language model (MLM) objective, is a highly effective training technique used in the transformer's encoder (Devlin et al. 2019). At its core, the method involves randomly "masking" or obscuring a fraction of the input tokens. This compels the model to correctly guess the masked words based on their surrounding context. MLM encourages the model to develop a robust understanding of the sentence structure and contextual associations among words.

Let's consider an example. Suppose the sentence is, "He went to the market." During the training, a percentage of the words – say "went" and "market" – might be hidden or replaced. The model must then predict these words using the unmasked context, "He ___ to the ___." The model is thus forced to understand the essence of the sentence to accurately predict the masked words. The predicted words are then checked against the original words, and the model is punished or rewarded based on its predictions. Over time and repetition, the model becomes more adept at contextual understanding and prediction.

The brilliance of masked token training becomes evident in tasks like next-sentence prediction and language translation. The encoder's comprehension of the sentence generated by masked training allows it to effectively capture the context and generate suitable translations or predict the next sentence. It's vital to note that the percentage of hidden words during training can vary, but it's usually set around 15% for models like BERT (Devlin et al. 2019). This fine balance ensures the model is challenged without being overwhelmed, thereby ensuring consistent improvement in language understanding.

On the other hand, encoder-decoder paired training employs two separate models. The encoder captures the contextual representation of the input, transferring it to the decoder that generates the output. The encoder-decoder paired training is a fundamental aspect of sequence-to-sequence (Seq2Seq) models, involving both components, the encoder and the decoder. They work collaboratively in a multistep process that not only ingests information but also transforms it into comprehensible output. This technique is particularly beneficial in tasks like machine translation, where input (a sentence in one language) needs to be transformed into a completely different output (the same sentence in another language).

In the first half of this process, the encoder digests the input sequence, learning to distill the important information from it, and transform it into an abstract context vector representation as we have seen. Essentially, the encoder parses through the sequence to extract and encode the context into the dense vector that should ideally capture the essence of the input. After this, the context vector serves as the bridge that connects the encoder and the decoder, carrying meaningful information from the former to the latter.

In the second half, the decoder takes the baton by receiving the context vector. It then leverages this information to generate the output sequence. The decoder recursively generates the output, predicting the next token based on the context vector and previously generated tokens. During the training process, the model utilizes a technique called *teacher forcing*, where the actual output sequence (from training data) is used as the input for the next-time step, rather than the previously predicted output. This way, the model learns from its errors promptly and realigns its predictions with the correct sequence, effectively amplifying the training's efficiency. As a cumulative result, this encoder-decoder training accosts models with the acuity required for complex language tasks.

Training Time

The training of LLM transformers such as GPT-4, PaLM, or Gemini is a time-intensive task that may range from several days to months, depending on the scope of the model, computational power, and the size of the training data. OpenAI, which developed the GPT-3 model, disclosed that training this 175-billion-parameter model on a machine with a single graphics processing unit (GPU) would take approximately 355 years (Li C., June 3, 2020). Although this time is drastically reduced by parallelizing the training across multiple powerful machines, it would still take several weeks or even months on clusters of graphics cards. Notably, this time frame doesn't include the crucial pre-processing and tokenization of training data, which can also take a substantial amount of time. Beyond the time involvement, training LLMs is also energy-intensive and incurs significant financial costs.

Transfer Learning

Despite the intense training commitments, transfer learning provides a convenient shortcut. With this method, a pretrained model, such as GPT or BERT, is fine-tuned on a specific task. Because these models have already learned a broad array of language features, they can efficiently adapt to a new task with less data and training time.

In the context of LLMs, transfer learning is particularly significant for two reasons. First, training LLMs from scratch is a resource-intensive endeavor, requiring substantial computational power, time, and specific expertise. The use of pre-trained models that have already learned fundamental language features alleviates this burden, making AI more accessible. Second, transfer learning allows for model optimization even with smaller amounts of data for specific tasks. As these LLMs have acquired a general understanding of language from their extensive training, they only need a smaller dataset to adapt to a specific task, say for instance, medical text generation or legal document analysis. Thus, transfer learning forms a crucial bridge between broad language understanding and specialized applications.

Model Alignment

Since these large LLMs are essentially trained on a large collection of internet text to predict the next token of a sequence, they may not be very good at correctly understanding and returning what it is that the user wants from them, and may produce logical mistakes and hallucinations. That is to say that these models aren't aligned with the requirements of the user. So how does one take a LLM that through its training has an understanding of language, concepts, and even code and align it to respond meaningfully to a user's questions?

This is done via techniques called Reinforcement Learning from Human Feedback (RLHF) (https://openai.com/index/instruction-following/; https://openai.com/index/fine-tuning-gpt-2/). The idea is to use feedback provided by human users to get the model to learn what is expected of it. The first step is collecting a dataset of demonstrations of instruction-response pairs that are written by humans. Think of this as a dataset of prompts and a demonstration by a human of what the expected output should be. This dataset is first used to fine-tune the pretrained LLM.

Once this is done, the LLM has a baseline understanding of instructions. Next, for a dataset of prompts, the model is made to generate multiple outputs. These outputs are then compared and rank-ordered by human labelers based on alignment guidelines. The data can be used to train a reward model (RM), which is an independent model whose only job is to predict which output from the LLM a human labeler would have preferred. This RM is key, as it is used as a reward function to further fine-tune the LLM.

At the end of this alignment process, the LLM responds in a more meaningful manner to being prompted and is helpful to users and aligned with their intentions. However, one of the challenges in doing so is that aligning a model on specific types of tasks (the sort of things most of us ask these models for) can make their performance worse on things like academic or programming tasks (https://openai.com/index/instruction-following/).

Process Supervision

What we have described so far is *outcome supervision*, meaning that feedback is provided only on the final outcome to the prompt. However, there is refinement used to further mitigate hallucinations, specifically on logical problems. This is *process supervision*, in which feedback is provided at each step in a chain-of-thought output (https://openai.com/index/improving-mathematical-reasoning-with-process-supervision/). This is similar to the alignment process described above, except that the RM rewards the model for following an aligned chain-of-thought; each step of the model's thinking is checked against a human-approved process.

This applies in particular to problems that need to be solved following a complex chain of reasoning like math or coding problems. The model generates a step-by-step output of its working, and each of these "working" steps are subjected to the RLFH process. It has been seen that this supervised alignment process results in the LLM performing better than an outcome aligned LLM across the board on solving math problems (https://openai.com/index/improving-mathematical-reasoning-with-process-supervision/).

Multimodality

We have thus far focused on LLMs, and the revolutionary design structure of transformer models as applied to text. However, one of the key features of transformer models is their ability to accept diverse types of input data events, not just sequences of text. This is due to the model's architecture, which is designed to process input data as a sequence of vectors, irrespective of the type of data being handled.

Consider the prospect of building image-to-text models. Conventionally, CNNs have ruled the realm of image recognition and processing. But developments like the vision transformer model (Dosovitskiy et al. 2020) have demonstrated that images can be effectively vectorized and fed as input to transformer models. In this scenario, an image is divided into a sequence of fixed-size nonoverlapping patches, which are then linearly embedded into vectors to be processed by the transformer-based architecture. The model functions as an image encoder and a text decoder, effectively transforming an image into a natural language description.

The beauty of transformers is that they can also work vice versa, creating text-to-image models. These models would involve a text encoder and an image decoder. The text input is transformed into a dense representation that encapsulates complex contextual and semantic information. Subsequently, this information is used by the image decoder to generate corresponding images or augment existing ones. Successful examples in literature include models like DALL-E from Open AI, which generates highly coherent images from textual descriptions (OpenAI 2021; Ramesh et al. 2022).

DALL-E's vocabulary has tokens for both text and image concepts. The first iteration of DALL-E was trained to accept both text and image as a single stream of data, where each image caption is represented using a maximum of 256 tokens with a vocabulary size of 16,384, and the image is represented using 1,024 tokens with a vocabulary size of 8,192. This allows DALL-E to not only generate an image from text but also modify or regenerate any rectangular region of an existing image (OpenAI 2021).

Adapting transformers to different data types also expands into arenas like video processing (Selva et al. 2023), sound synthesis (Verma & Chafe 2021), protein sequence modeling (Chandra et al. 2023), and more. This makes these models not just limited to encoding sequences of words but of any data that can be represented as a sequence of vectors.

The ability to process diverse types of input data is one of the major qualities that makes transformers such a groundbreaking approach in machine learning and artificial intelligence. It has unleashed a new realm of possibilities, ranging from text, image, video, and even biological data processing, igniting a plethora of opportunities for cross-modal and transdisciplinary innovations in the field.

Artificial General Intelligence?

Large language models (LLMs) such as GPT-4, Gemini, PaLM represent a significant advancement in AI, moving us closer to the goal of artificial general intelligence (AGI).

An early 2023 paper by a Microsoft research group shows that GPT4's performance is strikingly close to human level performance, and that given these capabilities it could be viewed as an early version of AGI (Bubeck et al. 2023). These models, built on neural networks and trained on extensive data, demonstrate an ability to understand and generate human-like language. In the context of AGI, this means an AI that can generalize learning and adapt to a variety of tasks, much like humans.

The versatility of LLMs is best illustrated by the performance of such LLMs across varied tasks. For example, GPT-4 scored in the top 10% of human test-takers on a simulated bar exam, a notable jump from GPT-3's performance (https://openai.com/index/gpt-4/). This kind of adaptability, moving from legal reasoning to creative writing or technical coding without task-specific tuning, aligns with AGI's foundational principle: the ability to apply general intelligence flexibly across domains.

In creative and problem-solving tasks, LLMs have shown remarkable capabilities. GPT-4, for instance, demonstrated the ability to generate novel medical hypotheses, suggesting its potential use in accelerating research (https://openai.com/index/gpt-4/). GPT-4's proficiency across a spectrum of tasks showcases elements of generalized intelligence. For example, its ability to perform well on standardized academic tests (like the SAT or LSAT) without specific preparation indicates a level of understanding and reasoning that transcends specialized training (https://openai.com/index/gpt-4-research/). This mirrors human cognitive flexibility and suggests a form of AI that can apply its "knowledge" in varied contexts, a key characteristic of AGI. In creativity, LLMs have been used to compose music and write poetry, activities traditionally considered the domain of human intelligence.

Even in the hallowed realm of pure mathematics, LLMs are going beyond what is known by mathematicians and computer scientists (Castelvecchi D., 2023). Google's DeepMind reportedly helped solve a long-standing mathematical puzzle (Castelvecchi D., 2023).

The road to AGI requires addressing current limitations of LLMs. This includes enhancing their sensory understanding and physical interaction capabilities. Ethical considerations also play a crucial role in the development of AI, focusing on the implications of model outputs and data biases (Bender et al. 2021; Gallegos et al. 2024; Kotek et al. 2023; Acerbi & Stubbersfield 2023). As LLMs evolve, their integration with other AI forms, like robotics and sensor-based AI, will be key in realizing a more comprehensive form of AGI.

This chapter has taken you through the architecture and the innermost workings of GenAi.

With this chapter, you have learned the basics of GenAi. And you are now equipped to move forward to greater knowledge, especially the combination of advanced neuroscience with AI. It is clear how the architecture of GenAi is indeed inspired by how humans think. However, if we want speed, we don't build a longer pair of legs, we build wheels and an engine. That is what GenAi has done – taken the principles of motion and locomotion to an advanced state.

Just as previously we all had to learn an entirely new technology – the digital world – we now are challenged to learn the latest.

The jargon of GenAi – tokens, encoders and decoders, and more – will become ever more integrated in our professional and personal lives. It will be AI as a second language. The primates of Silicon Valley have given us all a new language.

CHAPTER 2

The Perceptual Brain

Our senses are the foundation and the very essence of our life.
We rely on them for literally *everything*.

Taste, touch, scent, sight, and hearing.

The five senses. Individually and collectively, they power our perceptions, enhance our understanding of our existence, and bestow the blessings of enjoyment upon us.

They are deeply rooted in our earliest cave-dwelling days because they were essential to survival. Today, we have swapped that dependence out for the pleasures of great food, art, music, and more.

This chapter will take you on a fascinating journey among the senses, through the portal of neuroscience. And you will learn about the many and diverse ways that GenAi is revolutionizing so many aspects of our sensual perceptions today, from perfumes to pop music, compelling imagery to personal identity.

To be effective, the GenAi algorithms must utilize memory structures and a deep understanding of consumer behavior in order to construct captivating sensorials that elicit strong emotional responses from the intended recipients. By utilizing generative principles that have their origins in neuroscience, these algorithms generate fragrances that contain components of cultural, ritualistic, and temporal significance that are specific to each individual consumer. In order to exert maximum influence, the imagery generated is specifically customized to align with the consumer's favored locations, occasions, and times of day. In the same way that recipes and taste profiles incorporate nonconscious influences and ingredients that reflect the consumer's specific geographic and cultural heritage, so too are music and sonic signatures skillfully crafted to correspond with the consumer's preferences.

GenAi democratizes the creation of sensory pleasures. No longer do you have to rely on arcane expertise, or "license" formulas from cartels that control most of the world's market for fragrance and flavor. Intelligently using memory structures, and understanding the neuroscience behind sensory perception unleashes the inner creativity contained within every human. While fragrance houses and flavor houses obscure their path to fragrance, GenAi powered by neuroscience can make transparent the design path to a fragrance or a flavor.

Within each of us, locked in our hearts is a great musician – and we have been told that either through a lack of talent, or a lack of discipline, or the lack of fun and easy training we have never been able to unlock and unleash the musical genius within. We have longed to draw, to paint, and to create meta magical themes in concept and color. We have never been able to do that. The chef and mixologist within us has wanted to create cocktails and Michelin star dishes with ease, and we have told ourselves that we had neither the talent nor tools to make that magic in the mouth happen. How many times have we looked at Chanel, or Jo Malone, or Fredric Malle and said, "I can do better," only to be told that the expertise for perfumes is landlocked and confined to a few cities alone. Everything in us rebelled against this "artistic injustice." GenAi unlocks the artist within. So every human can unleash the sensory master within. This unlocking requires us to infuse GenAi with the neuroscientific understanding of art and craft. Mozart never needed neuroscientists, and Picasso dispensed with them easily enough, because they intuitively understood the neuroscience that underlies masterpiece creation. Powering GenAi algorithms with these neuroscientific learning makes every human a master, and every creation a masterpiece.

Come along for the ride.

The Scent of Success

The olfactory system is a silent influencer, a hidden persuader in the human experience. Our sense of smell is intimately linked to memory and emotion, residing within the brain's limbic system. Science has shown that olfactory bulbs act as the brain's aroma designers, crafting fragrant memories and sensations that linger far longer than visual or auditory impressions (Maaike et al. 2018). What we smell can transport us back in time or influence our behavior, making it a potent force for any marketer who knows how to harness its effects. A newborn child literally smells its way to nutrition, food, and the warmth of care. This sense is formed so wonderfully well at birth.

Take a moment and imagine walking into a bakery with the warm, yeasty aroma of fresh bread wafting in the air. This scent likely ignites a cascade of memories and emotions. The olfactory bulbs have a direct line to the amygdala and hippocampus – our brain's emotion and memory centers. It's not just bread; it's a whiff of nostalgia, a taste of a simpler time. By infusing their space or product with an evocative scent, a marketer can leverage these neural connections to create a strong brand identity.

The Neural Pathways of Scent

The journey of a scent starts with the inhalation of odor molecules, which interact with the cilia of the olfactory receptors. These cells transform chemical stimuli into electrical signals. The signals are then carried by the olfactory nerve directly to the olfactory bulbs, actuating a neural choreography in the brain that awakens memories and emotions. The short synaptic path from nose to brain underscores the primal potency of scent in our lives.

The sense of smell is perhaps the one that can most powerfully affect our emotions and behaviors. Helen Keller, who experienced the world from the perspective of someone who was both deaf and blind, called smell the "fallen angel" of the senses (Keller 1927). Yet her recognition of the power of olfaction aligns with modern neuroscience. Despite constituting only about 1% of our brain, our olfactory system punches far above its weight in emotional and memory impact. It is this sense of smell that our early ancestor used to find foods to favor, and foods to avoid. Olfaction became a powerful tool for survival itself.

How Smell Shapes Experience
Our nasal passages are lined with millions of olfactory receptors that recognize and bind odor molecules. When activated, these receptors relay information to the olfactory bulbs. From there, the olfactory information is dispersed to the amygdala and hippocampus, which contribute to our emotional responses and the consolidation of memories. The olfactory bulbs have extensive connections within the brain, underscoring how deeply entwined scent is with our perceptions and actions. Smell is contextual – the same scent, when presented in the context of aging cheese is pleasurable, but when presented in the context of dirty socks is abhorrent. So perfumers take care to present the context to interpret smell. GenAi algorithms should learn well that fragrance is highly contextual.

The Pheromone Phenomenon
Our ability to smell plays a critical role beyond environmental interactions – it's also foundational in interpersonal communication. Pheromones, the chemical-signal messengers secreted to affect the behavior of others, are detected by our olfactory system. These subtle scents can influence attraction and social cues on an almost subconscious level. Marketers can capitalize on this knowledge by choosing brand scents that evoke the right feelings and reactions from customers, drawing them in as though by instinct. Fragrance shapes, in powerful ways, evolutionary selection, by automatically identifying partners who might be biologically compatible with us through preferred fragrance.

The Potency of Scent Memories
Scent is the sense that's most closely linked with our memory, specifically our episodic memory – "the mental time machine" of who, what, when, and where. It's the aroma of your grandmother's home-cooked meals, the tang of the ocean at your first beach holiday. These olfactory experiences are engraved in our memories with precise details, robustly more than any other sense. Marketers have the opportunity to create memorable brand experiences by using scents that evoke these personal and affective memories in consumers (Rimkute et al. 2016).

Just as a bad smell can drive customers away, a signature scent can draw them in and have a tremendous impact on sales and brand loyalty. If well-executed, a scent can distinguish a brand so poignantly in a consumer's mind that even the whiff of a similar aroma can recall the pleasures associated with the brand.

Scent and the Dynamics of Desire
Perfume ads often portray scent as a pathway to desire, a not-so-subtle nod to our olfactory system's role in attraction. While scent marketing isn't about seducing consumers literally, it is about evoking the desirable emotions and memories tied to our evolutionary background – comfort, safety, appetite – that unconsciously drive decision-making.

In the world of perception, *smell* holds a subtle yet profound power over our emotions and behaviors. Unlike our other senses, which are mediated via intricate pathways in the brain, the olfactory system enjoys a privileged direct line to the neural realms of emotion and memory. This direct neural expressway grants fragrances a potent ability to evoke vivid recollections and emotional states, a phenomenon deeply rooted in the architecture of our brains.

Consider how scent can act as both a time capsule and a trigger. From the fragrance of a first-loved perfume to the comforting smell of freshly baked cookies reminiscent of childhood, olfactory cues are no mere passive players. They exert a profound influence that marketers can masterfully exploit. The anatomy of our scent processing apparatus mirrors our evolutionary heritage, where smell was a crucial ally in navigating the environment.

Nature's Subtle Social Network
While we don't rely heavily on pheromones as some animals do, our olfactory system still plays a role in social dynamics and interpersonal connections. Subconsciously, we are influenced by the scents around us and attracted to or repelled by them, aligning our behaviors with these olfactory cues. This highlights the importance for marketers to understand the aromas associated with their brands and how they impact customer perception.

Scent becomes even more intriguing when considered in light of epigenetic modifications – changes in gene expression without altering the DNA sequence. Researchers have observed that olfactory experiences can lead to epigenetic changes, adding another layer to how smells can shape an individual's predisposition toward certain behaviors, including brand preferences (Dias et al. 2014).

Immortalizing Brand Aromas
Olfactory cues are etched in the memory with great fidelity, which suggests that a thoughtfully chosen brand scent has the power to transcend time and context in customers' memories. Savvy marketers can create signature scents that become synonymous with their brand, embedding within their customers an emotional attachment that endures and drives loyalty (Batat 2019).

In the bustling aisles of commerce, strategic deployment of scent has been shown to improve not just mood but also sales metrics. The presence of congruent ambient scents has been linked to increased dwell time and spending in retail environments, showcasing the tangible benefits of olfactory marketing. This delivers a message loud and clear: Never underestimate the potential return on investment in the realm of aromatic appeal.

When considering olfactory marketing, subtlety and brand alignment are key. The chosen fragrance should evoke the essence of the brand without overpowering the sensory stage. This application of scented finesse can transform shopping from a mundane task into a delightful sensory journey, paving the path for enhanced customer engagement and brand sentiment.

The Aroma of Innovation   The marketing world is ripe for innovation through olfactory branding. Technology now allows brands to engineer custom scents that can be deployed and diffused in highly controlled ways, allowing for a scent-scape that evolves with customer interactions and even time of day. This brings an added dimension to the concept of experiential marketing, turning it into a holistic encounter that activates all senses and potentially unlocks deeper connections with consumers.

Navigating the Fragrant Marketplace   When marketing through olfaction, subtlety is key. A scent should whisper, not shout, gently guiding emotions and decisions rather than overpowering them (De Luca & Botelho 2021). Knowledgeable marketers can exploit this discreet yet compelling influence to craft an environment or product that becomes a multisensory experience, inviting loyalty and positive associations one breath at a time.

Emotions and memories are closely associated with the sense of olfaction. So when using GenAi to design perfumes and fragrance, it is important to trace a transparent path of memory structures that link the consumer to the constructed fragrance. This path becomes different from country to country and culture to culture as memory structures are different. It truly is inconceivable that a set of designers sitting in a country, without the knowledge of memory structures can envision, or dictate what someone else in a different country and culture and with an entirely different set of memory structures can desire. GenAi algorithms can bridge the gap by extracting memory structures specific to a country and culture.

The Eyes Have It

Let's consider vision, a sense that dominates our perception of the world. Neuroscientific research shows that approximately one-fourth of the human brain is dedicated to visual processing, indicating its primal importance to our survival. Our proclivity for sight over other senses also finds roots in our evolution, which favored visual acuity as a means to better interact with and interpret our environment.

Visual Dominance   With 70% of our sensory receptors nestled in our eyes, human beings are, above all, visual creatures. The impact of this on the marketplace is noteworthy; a product's visual appeal can make or break its success. Recognizing this, marketers must strategically integrate compelling visuals to capture and sustain consumer attention (o'g'li & Zamonbekovich 2023). A colorful display or an engaging visual story can make the difference between a product that resonates and one that is overlooked.

Visual Workshop   Interestingly, the act of seeing is less about our eyes than the brain behind them. Our eyes function as biological cameras, but it is the brain that translates light into the rich mixture of colors, shapes, and motion that we experience. The significance for marketing lies in creating visual content that is not just seen but meaningfully processed and remembered, crafting an immersive experience for the consumer.

A Colorful World As the artist Paul Cézanne observed, the perception of color is a construct of the brain, not an inherent property of objects. Our brain assigns colors based on the way light reflects off objects. This suggests the strategic use of color can play a pivotal role in how a product is perceived and can influence consumer behavior significantly.

The Evolutionary Advantage of Two-Legged Sight Our ancestors' transition from four-legged to bipedal locomotion granted us a panoramic view of the horizon, critical for survival on the savannah. This elevational advantage translates directly to consumer spaces: arranging products at the optimal height and angle can harness the natural propensity of customers to scan their environment efficiently, contributing to improved sales.

A Predator's Gaze We share with our predatory ancestors' eyes positioned to best track and engage with prey, a trait we bring unconsciously into interpersonal interactions and consumer decisions. This evolutionary heritage underpins the importance of direct eye contact in marketing, fostering a more compelling and persuasive connection with potential customers.

The Marketplace and Arresting Attention Our predilection for attractive visuals isn't restricted to the natural world – it permeates the marketplace. Product packaging, branding, and advertising campaigns that utilize visually appealing imagery are more likely to resonate with consumers, as our neural pathways are primed to respond to aesthetically pleasing sights. Be it a package in a crowded aisle, or a billboard that zips by on a highway, or a carousel of product imagery that fights to stand apart in a crowded online superhighway, arresting attention through visual appeal is critical. GenAi algorithms that generate visual imagery should learn from what arrests visual attention in the brain.

The Authentic as Aesthetic The importance of aesthetic appeal extends to societal interactions, with implications for marketers. Attractive individuals are often perceived more positively, influencing everything from hiring decisions to social treatment. In this light, featuring visually appealing people and scenarios in marketing can unconsciously bias consumers toward a favorable perception of a product. It begs the foundational societal question: "Aesthetic according to whose standards?" There is an increasing demand from consumers to show authenticity as the new aesthetic. To eschew old fashioned ideas of beauty and aesthetic, but to reveal real and regular people doing everyday things as the new aesthetic. Not showing slender models walking on red carpets, but showing tired parents cleaning red stains out of a carpet. Society is still searching for answers, to these vexing puzzles, and the brain views cognitive fluency as another definition of aesthetic – simply, if it is simple, it is elegantly beautiful. GenAi algorithms that understand these nuances will create imagery that is aesthetic in crowded marketplaces.

Balancing Narrative and Visuals While the power of visuals is uncontested, this does not negate the importance of a compelling narrative. The interplay between story and product must be delicately orchestrated to maintain consumer engagement. Rather than abrupt transitions into product features, we must weave product details into the fabric of the visual and narrative tapestry, ensuring a seamless consumer experience that engages both the brain's emotional and analytical spheres.

Rules of how the brain perceives visual form and interprets it should guide GenAi algorithms that create art and visuals. In the chapter on packaging, we will reveal and study the many findings of neuroscience that inform how visual imagery is processed in the brain. This scientific understanding helps us appreciate the science behind Escher, Magritte, and the impressionists. These rules can then be explicitly included in the guiding frameworks of GenAi algorithms that will then generate imagery that uses these principles.

The Flavor Matrix

The Anatomy of Taste At the center of taste perception is the tongue, a muscular organ swathed in mucous and studded with taste buds that flirt with flavors as they move across our palates. Each taste bud is a microscopic bastion of gustatory cells that translate chemical compounds into the sweet, salty, bitter, sour, and umami tastes we experience. This translation is not simply about enjoying food; it's about evaluating our food sources for nutrition and potential toxicity.

Compounds convey their flavor messages via receptors on taste cells, initiating a cascade of neural signals. Taste receptors are selective, each tuned to specific flavor molecules – sweet receptors, for instance, wait for sugars or artificial sweeteners, while umami receptors respond at the touch of amino acids. These signals are carried away by gustatory nerves to our brainstem, thalamus, and eventually, the cerebral cortex, translating chemistry into conscious taste perception.

The story of taste buds is more than a mere sensory narrative; it's survivalist prose. Bitterness often signals toxins, prompting an immediate, instinctive rejection – a built-in aversion therapy. Sweetness, on the other hand, signifies energy-rich sources. These elemental tastes form a matrix of flavor preferences and aversions that influence our dietary choices and, by extension, our health.

A Union of Senses Taste and smell are close relatives in the sensorial family, each enhancing the other's information to create a unified perception of flavor. Smell often captures our attention first. A whiff of roasting garlic can trigger salivation and anticipation of a flavor yet to touch the tongue. A symbiotic relationship indeed exists, where the loss of one – often smell – can produce flavor sensations exclusively in shades of bland.

Chemical Drivers of Desire   While we savor the euphoria induced by a scoop of gelato or a crunchy potato chip, behind the curtain, neurochemicals are being orchestrated by our dietary choices. Foods high in salt, fat, or sugar play on our neurotransmitters, releasing endorphins and dopamine, those pleasure-inducing chemicals that not only make us feel good but might also have us coming back for encore performances.

Carbohydrates hold a unique sway over our mood, as their intake boosts serotonin levels – a neurotransmitter associated with calm and well-being. This biochemical tango may underscore the cravings we develop, potentially explaining why some reach for the macaroni and cheese when clouds loom large on emotional horizons. Intriguingly, studies suggest that these cravings could be choreographed by our genetic makeup, adding another spicy layer to the narrative.

Taste Transcends   So how does one activate taste effectively? Mirror neuron systems activate at the sight of someone relishing that triple chocolate cake, seducing the viewer into a phantom sensory experience. Brands that can stimulate vicarious taste through sight or even thought tap into a powerful transactional potential that turns whims into wallets (Lee & Spence 2022).

A depiction of a foamy cappuccino can convey more than a verbose menu ever could. However, authenticity is key – artificial props are the false notes, breaking the spell and dampening the appetites that might have been brewing. Marketers must craft experiences that ring true, delivering the promise of flavor to the consumer.

GenAi algorithms need to rely on the science of taste – molecular resonance that happens when compounds that stimulate the gustatory complex overlap in spices and foods. Chefs and mixology maestros intuitively know that, and they know what to add and how to blend them perfectly. The world's recipes offer this tantalizing bit of evidence that facilitates building transformers to make this knowledge easily available to the casual user.

The Tactile Touchpoint

Touch is our most fundamental sense, the first to develop in utero and the most essential to our survival and emotional well-being. This primordial connection remains with us throughout life, cementing bonds, soothing distress, and shaping our interactions with the world and the objects within it. The skin, enveloping our body in a continuous sheath, serves as a canvas for sensation, with every brush, pressure, and texture telling a tale to our perceptive minds. A majority of the homunculus in the brain is focused on sensations from palms and lips – the primacy of touch.

The Embodiment of Emotion   Skin-to-skin contact releases oxytocin, the "love hormone," which fosters connection, trust, and a sense of safety. This is biologically ingrained in us – infants instinctively seek the comfort of their mother's touch for reassurance. A touch can assuage anxiety and convey compassion without the need for words, an unspoken language that is universally understood and deeply rooted in our neural networks.

Touch's significance goes beyond emotional interplay; it underpins our sensory perception of the world. The tactile sensory system is distributed throughout the body but is most concentrated in areas like the hands and face. The density of mechanoreceptors in these areas makes them exquisite instruments for perceiving the world's physicality – far surpassing the capabilities of less-sensitive body parts.

Sense of Self and Surroundings   The brain's somatosensory cortex processes touch signals, turning physical stimuli into perceptions of texture, temperature, and pain. This cortical region is proportionally mapped to the sensitivity of different body areas, with hands and fingers claiming a significant expanse – befitting their exploratory role in the human experience.

Physiology of Perception   Our tactile discernment relies on various specialized receptors: Meissner's corpuscles register light touch, Merkel's discs sense pressure, Ruffini endings detect skin stretch, and Pacinian corpuscles respond to vibrations. Each type contributes to the rich tapestry of touch sensation, fast-tracking messages through the nervous system to alert us of environmental changes or potential harm.

Sensing Pain and Pleasure   Pain, a crucial sensation for survival, is the body's alarm system, signaling injury and danger. Touch receptors like nociceptors are quick to respond to harmful stimuli, prompting immediate and protective reactions. But pleasure and comfort are the other side of touch's coin, mediated through gentle, calming stimuli that signal safety and promote bonding. The old Lamisil advertisement for a remedy for toenail fungus showing a toe nail being gently lifted and ripped from the toe triggers pain reactions in the brain. This shows the power of both imagery and the dominance of the sense of touch.

Tactile Tactics   Understanding the consumer's tactile expectations is just as crucial as acknowledging visual appeal or auditory branding. The quality of product materials, the weight and balance of an object, even the shape and smoothness of a container's surface can subtly influence consumer behavior. These tactile cues can quietly drive decisions toward comfort, familiarity, and satisfaction (Rifqiya & Nasution 2016).

From the fleshy softness of a ripe peach to the cool sleekness of a smartphone, texture speaks, and savvy marketers are listening. Products designed with a conscious engagement of touch can lead to enhanced consumer-product relationships, as tactile feedback is incorporated into product memory and experiences.

How might one craft an appealing tactile experience in the marketplace? Products that grace the palm should be sculpted for pleasure, their contours and textures resonating with deft fingertips. Merchandise should be welcoming to the touch. A cozy sweater that wishes to be hugged, a chair that promises comfort at first contact – all this speaks in the silent language listened to by the sensory-rich appendages we extend to the world.

Designing for Delight Envision a store where every interaction invites a tactile conversation – textiles beckon with soft whispers, gadgets impress with smooth finishes, and packaging crinkles delightfully under curious fingers. It is a world where the tactile dimension augments the visual – the shape, the resistance, the finish – all these elements unfold in layers of sensory feedback, guiding the narrative we co-author with objects we encounter.

In a marketplace where attention is now the new currency, a brand's touch can become its signature. The inviting softness of a luxury car's leather seat, the reassuring sturdiness of a high-quality tool, or the warm embrace of a perfectly fitted sweater – these are tactile messages that convey value, care, and desirability, directly engaging the consumer's sense of touch, the most personal of sensory experiences (Zuo et al. 2016).

GenAi algorithms need to incorporate the sense of touch in a few important consumer "touchpoints" – all product descriptions and innovations in their descriptive language need to include touch and tactile feelings. Design of products and visual imagery need to include texture as an important parameter and not as an afterthought. Product designs should pay special attention to texture, materials, temperature, roughness, and tactile sensations and associated emotions in both design and description.

Sounds of Silence

Listening forms the cornerstone of human interaction and environmental awareness. Our auditory system is exquisitely tuned to detect a vast array of sounds, from the minute rustling of leaves signaling a predator in the wild to the complex harmonies of music that stir human emotion. Understanding the science of hearing reveals how sound not only transmits information but also evokes affective responses, creating an auditory tapestry that resonates through our lives.

Decoding the Echoes of Evolution The process of hearing starts with sound waves traveling through the ear canal, implicating both the physical movement of particles and the subsequent translation into electrical signals within the brain. This conversion happens through the work of tiny hair cells in the cochlea that respond to different frequencies, a system honed by millions of years of evolution to favor communication and survival. From an evolutionary standpoint, the ability to hear has always been intertwined with the ability to connect and thrive.

The Art of Auditory Perception   Once sound information reaches the auditory cortex, it's not merely identified but also interpreted. This interpretation colors our perception with the hues of memory and emotion, casting sound as a powerful mnemonic device. A melody can evoke a flood of memories and vivid images, transporting the listener back in time to the moment of the song's significance – an effect potent not only to individuals but also to large groups unified by a shared cultural soundtrack.

Sound does more than just register in the brain; it sets a cascade of neurochemical reactions. Dopamine, a neurotransmitter associated with pleasure and reward, is released in response to enjoyable music, akin to the effects induced by food or sex. These neurochemical symphonies can explain why certain sounds can instantly uplift our mood or drive us to action – a fact not lost on those looking to engage audiences or consumers.

From Lullabies to Eulogies   Throughout the human experience, specific sounds and music mark the milestones of our existence. The lullaby that soothes a newborn, a couple's chosen love song, the celebratory fanfare of commencement, or the solemn tones at a funeral – each auditory event is embedded in cultural and personal significance, echoing through our neural circuits, and influencing behavior.

Soundscapes and the Human Emotional Experience   Every environment has its unique sonic identity – a cacophony at a market, the tranquil hum of a library, the rhythmic pulse of a nightclub. The background sounds in these environments are not mere noise but an integral part of the experience, shaping our perception, behavior, and even physiological responses, like heart rate or pupil dilation. Awareness of this auditory ambiance can be a powerful tool for anyone endeavoring to create an environment, whether for leisure or commerce. There is an intimate connection of sound to emotion. Algorithmically cracking this code can lead to accurately pairing sounds, and soundscapes with products and marketing.

Why We Crave the Sonic Sizzle   Sound is intimately tied to anticipation and reward. The fizz of a soda being opened, the crunch of a crisp chip, the clang of winning coins – each sound is a cue that triggers a learned association and an emotional response. The sizzle of food, in particular, has been linked to the expectation of taste and satisfaction, establishing a sensorial feedback loop that can influence choice and preference.

Building Auditory Bridges   Creating an effective neurological iconic signature (NIS) for a product often involves crafting distinct sounds that become synonymous with the brand. When these sounds echo the experiences of peak enjoyment, like the snap of a chocolate bar breaking, mirror neurons in the brain light up, mimicking the act of consumption and enhancing the desire to partake. This auditory branding is not simply about standing out; it's about resonating on a personal and emotional level with the consumer (Minsky et al. 2017).

Sound and Strategy

For marketers and designers, the strategic use of sound extends far beyond jingles or slogans. It includes careful curation of all auditory elements in a consumer experience, from the music in a commercial to the acoustic features of a shopping space (Spence et al. 2019). A harmonious alignment of brand values and auditory cues can increase memorability and foster an emotional connection with the brand. These soundscapes can turn a passive listener into an active participant in the brand narrative.

Melodies and Memory

Perhaps more than ever, in our visually crowded and noisy world, the sound of your brand can serve as a serene siren song, beckoning consumers to look beyond the visual and engage with the full sensory spectrum your product offers. Whether aiming to soothe, energize, or inspire, embracing the full octave range of auditory influence in marketing can yield positive results.

Transformer technology can facilitate the creation of original sonic branding that takes the emotional landscape of the brand and brings it to life through sound. GenAi algorithms can analyze music and sound and extract the emotional content in them second by second. This enables verification and validation that the chosen sound not only is distinctive but also ties emotionally to the attributes of the brand or the evolving ad.

Now that you have finished this chapter, you have not just a comprehensive overview of how our senses function, especially at the nonconscious level. You have a tutorial on how to apply the combination of neuroscience and GenAi to achieve extraordinary success in your work – and even in your personal life.

CHAPTER 3

The Teen Brain

They are tempestuous, occasionally trying, and they love to test limits.

They are teenagers.

Understanding them has confounded adults from time immemorial. But modern neuroscience has cracked the teenage code.

Recent knowledge of the brain has rapidly advanced, and the mysteries of the immature mind have been revealed.

For marketers, this can be a boon. But now, the transformative breakthrough that is AI brings even greater marketing potential.

In this chapter, you will learn not only the inner neuroscientific workings of the teen brain and how best to reach and appeal to it. You will learn how the unique combination of neuroscience with GenAi can take your efforts to new heights of effectiveness.

In a way, GenAi is a teenager – great capability and unrealized potential. It hallucinates and misbehaves at times, but it is worth caring for and loving. So this chapter is not only dedicated to understanding of the teen brain but is an introspective on GenAi itself.

Come along for an in-depth trip into the teenage brain.

Language of Emotions: Marketing to the Emotional Maelstrom of Teenage Brains

The unfolding of the human brain, particularly throughout the teenage years, offers a fascinating insight into human behavior and decision-making. During this critical period of development, the emotional seat of the brain, known as the limbic system, rapidly matures. This neurological precinct, ruled by the amygdala, develops earlier than the prefrontal cortex, which is responsible for logical processing and executive function (Casey et al. 2008). Simply put, don't try to reason with a teenager.

What does this brainy evolution mean for marketers? In the fluid and volatile environment of teen emotions, marketing approaches must adapt. While adolescents navigate the choppy waters of emotional development, their cognitive capabilities are not yet fully developed. Herein lies the conundrum: How do we communicate effectively to an audience more attuned to the resonance of feelings than the rigors of logic?

The answers might lie in the lexicon they favor. Emojis, those modern-day hieroglyphs, are in many ways the perfect encapsulation of emotional fluency in teen communication. They offer direct, affect-laden pathways to teen hearts and minds (Lu et al. 2016). A smiley face can be worth a thousand logical arguments when it comes to the emotionally attuned teen brain.

In marketing, this translates to a tactical pivot from traditional "reasons to believe" to "feelings to experience." Teenagers may be skeptical of why they should use a product based on its qualities and features alone. Instead, tapping into the emotional storyline associated with a product can lead to a much more visceral and compelling connection.

This shift from reason to emotion should not be considered a reduction in complexity – far from it. The science of emotional engagement is intricate and requires a keen understanding of neural development. Marketing strategies that align their narratives with the still-developing executive functions of the teen prefrontal cortex can fumble inadvertently. Emotionally laden messages, however, are more likely to resonate and be remembered due to the more mature state of the limbic system.

When crafting a message, one should consider the intensity of emotions during adolescence. High arousal positively influences the attention and retention of an advertisement. Considering how hypersensitivity to emotional stimuli is a hallmark of the teen brain, creating emotionally charged marketing campaigns can be particularly effective.

Appealing to teen sensibilities does not mean throwing logic out the window. Rather, it suggests that marketers and GenAi should weave logical arguments within a fabric of emotional narratives. Their brands' stories can build on the memory and emotion duo, which are deeply interconnected in the hippocampus – a brain structure at its developmental peak in teenage years.

What could these emotional narratives look like? They are the shared triumphs and tribulations of adolescence. They could be stories of empowerment, self-discovery, friendship, and adventure – all common emotional currents in the lives of teenagers. By connecting the product to these experiences, a brand can transcend the mundane and become a cherished ally in the teen's personal odyssey.

This emotional journey, however, must be navigated with finesse. Overloading the teen brain with exaggerated emotions or appearing disingenuous can backfire (Price 2017). Authenticity is the name of the game, and adolescents, despite – or perhaps due to – their emotional bias, are particularly astute at recognizing when they are being condescended to or manipulated.

Herd Mindset: Prefer Friends over Family

Adolescence is a time of significant neurological transformation, a period where the brain undergoes extensive remodeling, not unlike the renovation of a historical building – one where the past foundation must support state-of-the-art enhancements. Understanding these neurodevelopmental renovations is crucial for marketers who wish to create compelling campaigns resonating with teenage sensibilities.

The teenage limbic system undergoes swift maturation, sharpening the emotional acuity of young individuals. This internal tempest is sculpted by rapid synaptic pruning and myelination, which enhance the brain's affective circuits. Teenagers become more emotionally astute, but this astuteness often leads to a heightened sensation-seeking and risk-taking behavior that can shape their consumer habits (Cservenka et al. 2013).

The social brain hypothesis posits that humans have evolved to navigate complex social networks and that the adolescent brain is specifically tuned to social cues. This union of emotional vigor and social versatility offers a rich playground for marketers.

In the context of marketing, an understanding of the group dynamics driving adolescent behavior is critical. Adolescents exhibit a pronounced herd mentality, propelled by the evolutionary psychology of seeking safety in numbers (Kilner & Blakemore 2007). This deep-rooted tribalism can be harnessed by marketers who can position their brands as conduits to inclusion within desired peer groups.

It is within these social enclaves that trends emerge and propagate at a startling pace. Consequently, fads can dominate teenage purchasing patterns. Still, the underlying constant is the search for identity and belonging (Schwartz et al. 2005). For advertisers, ensuring that their products signify a badge of this coveted social capital can provide an edge in a market saturated by the noise of competing stimuli.

While the bonds of group belonging are powerful, within those groups a kaleidoscope of individualities strive for expression. The teenager's quest for self-discovery and personal expression is intertwined with their desire to fit in. This dual force can be leveraged by marketers crafting messages that applaud individuality while simultaneously reinforcing group identity.

Products and marketing narratives that embody authenticity are particularly successful with this demographic. Teenagers, with their innate ability to smell insincerity from a mile away, gravitate toward brands that they perceive as genuine and trustworthy. Messages that are honest, direct, and echo the real-world experiences of adolescents can create resonating bonds.

Furthermore, the advent of social media has amplified the potential impact of peer influence on teen decision-making. When teenagers see their peers' endorsing products or behaving in particular ways online, they are more likely to follow suit due to the mirroring of virtual and actual social networks. Tapping into the influential power of peer figures, therefore, can be a potent tool in a marketer's arsenal.

The branding of experiences rather than the selling of mere products is a subtle alchemy that can work wonders on the teenage brain. Creating an aspirational landscape wherein teenagers can project their fresh identities can create lasting associations with a brand. Thus, not just selling a shoe but selling the journey one undertakes with those shoes can make all the difference.

In the end, the teenage market is dynamic, complex, and profoundly human. Marketers who approach it with a blend of scientific understanding and empathetic engagement are likely to see their efforts bear fruit. GenAi must create emotional backdrops and compelling stories along which rational narratives may be provided. Algorithms must understand and capitalize on the *herd mindset,* something so well understood in the world of venture capital.

Almost Mature: Sensual and Yet Insecure

The landscape of adolescence is marked by a profound milestone: the surge of reproductive hormones that awaken a newfound sensuality and sexual maturation. During this time, the teenage brain is being sculpted by a torrent of endocrine changes, which critically imprint on the socioemotional circuitry of the brain. This hormonal revolution brings with it an intensified desire for peer acceptance, camaraderie, and romantic connections, highlighting sensuality as a new priority in their evolving identities.

The activation of the hypothalamic-pituitary-gonadal axis is at the heart of these changes, driving not just physical growth but also playing a substantial role in the psychological and social development of teenagers. The neurobiological upsurge caused by sex hormones like testosterone and estrogen is evident in increased preoccupation with body image, which is perceived as a proxy for maturity and, by extension, sensuality. The teenage quest for autonomy and self-expression dovetails with this emergent sensual awareness, adding layers to their consumer behaviors.

Neuroimaging studies have revealed that regions associated with reward processing, particularly the nucleus accumbens, are hypersensitive in teenagers, influencing their responses to social and sensual stimuli (Foulkes & Blakemore 2016). Marketers can capitalize on this by framing products as bridges to potential romantic interludes or as tokens of mature sophistication. In doing so, they align their value propositions with the teen's neurological reward system that endows such objects with heightened significance.

Teenagers tend to be acutely attuned to the judgments of their peers, especially concerning attractiveness and maturity (Carr 2015). This is due, in part, to the social brain's development, with areas like the medial prefrontal cortex (important in social cognition), maturing into the early twenties. Consequently, societal constructs around beauty and maturity may significantly influence product appeal among the teenage demographic, as these constructs often symbolize a transition into adulthood.

The sense of maturity and attractiveness also feeds into the concept of social currency – a commodity of heightened value in teenage peer networks. Brands that manage to permeate these networks successfully do so by translating the product's utility into the currency of mature sensuality. This implied promise of elevated status and peer admiration can turn otherwise ordinary products into potent totems of adulthood.

A distinct visual aesthetic that channels a mature, sensuality-oriented appeal is another tool marketers can employ. The teenage brain's heightened receptivity to visual stimuli, especially those with a sensual undertone, can catalyze a strong brand connection. This blend of marketing is less about the cold sell and more about sculpting an image that aligns with the teenager's internal narrative of aspirational maturity.

Apart from engaging the visual senses, auditory cues in marketing also assume high importance in targeting the teen consumer. Music associated with maturity, romance, and popular culture could strike resonant chords, due to its ability to facilitate strong emotional reactions and memory associations – both areas where the teenage brain shows peak sensitivity and reactivity.

Despite the significant lean toward sensuality and maturity in marketing to teenagers, authenticity remains an undiminished priority. Overstated or inauthentic claims may trigger teenage skepticism and damage brand credibility. A nuanced approach recognizes the teen's emerging critical thinking skills – courtesy of their maturing prefrontal cortex – and balances appeal with believability.

To engage a demographic that's internationally connected, faster-paced, and more visually oriented than any before, it's important for marketers to utilize social media platforms as channels of influence. Here, endorsed messages gain value from the peer leaders who broadcast them, merging the realms of mature allure with peer acceptance.

GenAi has a delicate task here – creating nuanced language that is sensual, connecting with emergent romance and sexuality that is central to the teen brain while being respectful and compliant with laws and cultural norms. These tasks are, however, most suited for algorithmic execution. Each country and culture's norms become guidance and constraints for message crafting by GenAi.

Person in the Mirror: Search for Identity

Neuroscience unveils that the ages between 20 to 25 are not just about discovering the best coffee shops or enduring the most epic road trips. It's a time when the brain is busy at the tail end of its construction phase. The process of myelination, wherein neural pathways get their insulating "white matter" sheaths, can be thought of as the brain's final push toward a more efficient, connected, and faster-thinking adult self. This period coincides with the emergence of a stronger sense of self and personal identity – a crucial bit of news for anyone looking to appeal to the wallets of this demographic.

Identity formation at this stage is akin to the last coat of paint on a canvas; the art was there, but now it reaches its full richness and depth. Neuroscience explains that the synthesis of experiences, values, aspirations, and societal expectations coalesce during this period to help establish a robust, hopefully coherent, sense of who one is and one's place in the world.

The fluctuating identity market – let's call it "The Self Shop" – is prime real estate for marketers aiming to appeal to a crowd that's literally wiring itself for long-term brand loyalties. The young adult brain is branding its neurons with memories; what it encounters now sticks like gum on a hot sidewalk.

From a marketing perspective, the audience is unique. No longer impulsive teens and not yet fully predictable adults, young adults are crafting their personal storyboards. They're the heroes of their own sagas, often featuring independence, career paths, social affiliations, and aspirational lifestyles. The savvy marketer knows that campaigns should not just sell a product but become a part of these personal narratives.

Social influence still pulls some strings, but the puppet show of adolescence gives way to a more selective and discerning audience. The pronounced effect of peers transforms into a nuanced concert of influences, as exemplified by the maturation of the prefrontal cortex, the brain's executive decision maker. The confluence of social inputs against a backdrop of a more refined self-concept means that young adults are likely to align with brands that reflect their emerging self-image and life philosophy (Twenge 2017).

At heart, this process is about seeking and solidifying one's distinct place in the world, with signature tastes and affinities. Brands that demonstrate values aligned with this introspective process often forge deep, resonant connections. Your marketing shouldn't be the loud friend who can't read the room; it needs to be the cool mentor who quietly guides without stealing the show.

The way to a young adult's heart? Speak to their evolving identities, not their past selves. After all, they're in the serious business of self-creation. Creating campaigns with an element of self-discovery or self-affirmation, like a travel adventure that beckons the spirit of autonomy or a gadget that screams cutting-edge individualism, resonates with the crux of their developmental agenda.

Another card to play is the exclusivity angle. At a time when one's identity feels like a hand-crafted sculpture, what's better than the notion of a product that's as unique as they are? Customization and tailored experiences sing the siren's song to a brain lit up by the dopamine that buzzes with the special and the singular.

Perhaps the difference now is that the humor is drier, the irony richer. The young adult brain loves wit that skirts the edge of their matured understandings – chuckles that come from a sophisticated place. It is not slapstick or people tripping over banana peels; it's the smart and subtly layered laugh of the in-crowd.

By the end of this formative stretch, the brain's marathon of selfhood is hitting its stride. These young adults stand ready, with identities mostly in hand, prepared for the adult world and its myriad choices – including brand loyalty that could last a lifetime (Arnett 2007). They are neither clay to be molded nor stone to be chiseled; they're artworks, fresh from the kiln, eager and warm for the right setting.

GenAi algorithms must be the curators who understand the worth of this prized demographic, crafting campaigns that speak the dual language of self and style with an understated confidence. Identity needs to be exclusive, but the language needs to be inclusive – teen brains rebel against injustice and discrimination. GenAi designs should deliver on identity, exclusivity, and inclusivity.

Musical Imprinting: The Unforgettable Playlist of Youth

Have you ever wondered why certain songs have such strong, immediate appeal, or why particular chords summon a torrent of nostalgia? Research shows that the music we encounter during our formative years, particularly between the ages of 14 to 22, leaves a profound and distinct impression on our brain (Fu et al. 2024). During adolescence and young adulthood, the brain is plastic, undergoing significant development, and this includes the auditory cortex responsible for processing sound.

Our neural development coincides with a peak in emotional intensity. It's no secret that teenagers experience feelings more deeply and vividly. Music, an inherently emotional medium, becomes coupled with this crescendo of emotions, essentially soundtracking life's drama at a time when our brains are wiring up for the adult world of social and emotional complexity.

The teenage brain is like an emotional sponge, and music is among the most absorbable influences it encounters (Bargh 2017). When we experience the pleasure of listening to our favorite songs, that's the reward system – particularly the nucleus accumbens – doling out dopamine. This neurological reward mechanism reinforces the memories and emotions associated with the music of our youth, grounding them in the bedrock of our identity.

Moreover, memory formation at this age is incredibly robust, thanks to the hippocampus which is zesty with youthful vigor. The connection between the hippocampus and the regions responsible for auditory processing becomes strengthened during adolescence. Melodies and lyrics heard in this era tend to become encoded in our long-term memory, often emerging effortlessly after years or even decades.

The reminiscence bump, a phenomenon observed in autobiographical memory research, identifies that people over the age of 35 have enhanced recollection for events from adolescence and early adulthood, particularly for music (Belfi et al. 2021). This implies that the melodies etched into our brains during this period are particularly potent for triggering memories, plucking the strings of our nostalgia with pinpoint accuracy.

Transitioning now from the halls of science to the streets of marketing, how does this array of findings bode for strategies aimed at the older demographic? It translates to opportunity, where leveraging the sonic elements from their golden years can render a product not just heard, but felt. The wise marketer will cue the soundtrack of the consumer's youth to evoke a sense of authenticity, familiarity, and a bond that transcends transactional relationships.

It's all in the mix tape of marketing – blend the new with the treasured tunes of yesteryear. GenAi algorithms that skillfully sync their messaging with melodies from this key developmental stage can find their audience more engaged, weaving past affinities with present possibilities. This is the kind of algorithmic output that makes for happy hours, where a simple jingle can lead consumers to linger longer and smile at the thought of your brand – to note the importance of the teen years of 14–25 in forming deep and resonant connections in the brain that vibrate fresh even after years of passing. GenAi should fully leverage this.

Selfie Science: Obsessive about Me

The rise of the selfie culture is inherently tied to neural processes that underpin the teenage sensation of identity formation. In adolescence, intense neural sculpting occurs within regions responsible for self-referential thinking, such as the medial prefrontal cortex (mPFC), which is crucial for one's concept of themselves. Social media platforms, with their ability to present an idealized online persona, interface directly with the mPFC's computation of self-identity, offering a tangible representation of teens' evolving self-perceptions (Valkenburg & Piotrowski 2017).

From a developmental perspective, the adolescent brain is a neurological construction site. Synaptic pruning, a process of eliminating neurons and synapses that are less frequently used, coincides with a surge in social awareness and self-consciousness. This refining of neural networks ensures that adolescents achieve a more sophisticated sense of self, essential for navigating complex social landscapes. Marketers must recognize the inherent value teenagers place on self-expression platforms and opportunities for self-display.

Mirror neurons, which fire both when individuals act and when they observe the same action performed by others, play a role in the developmental tug-of-war between individual identity and social conformity. When a teenager observes a peer receiving positive reinforcement for a shared selfie, the mirror neuron system simulates this reward, endowing the act of selfie taking with vicarious pleasure. This gives marketers an avenue to promote products that encourage and enhance the shared experience of self-expression.

As teenagers seek peer validation, the reward circuitry within their brains is highly active. In particular, the nucleus accumbens, an area associated with reward processing, is notably responsive to social stimuli (Hikida et al. 2016). The inclusion of one's image in a social context, such as in a group selfie, acts as a neural reward stimulant, further intensifying the appeal of platforms and products that amplify these experiences and integrate them into an individual's social repertoire.

Transitioning into application, the research suggests that marketing tactics that enable spontaneous self-portraiture and connectivity with friends can tap directly into this major cognitive shift. Products like selfie sticks or social apps designed to amplify image-sharing capabilities can be seen as tools that support the teenager's vigorous construction of self-identity. They resonate with the neural progression toward a unique, socialized personal identity strengthened by public acknowledgment.

Many teenagers' postings of selfies can be correlated to, essentially, a real-time curation of their ongoing autobiography – a pursuit intimately wired to the notions of memory and social cognition. Digital platforms that facilitate this process, along with filters and editing software, act as a canvas for teens to shape their ever-evolving story. Savvy marketers could create campaigns that showcase how their products live within these personal narratives.

Self-deprecation and wit can play well with the teen audience, provided they are carefully balanced against the backdrop of reinforcing positive self-image. Products that can both amuse and offer inclusion into a desirable identity construct are likely to capture the hearts of this demographic.

In conclusion, the obsession with capturing and sharing images of oneself is not mere vanity – it is rooted in the neurological development unique to the teenage years. GenAi that is aligned with this phase of brain development – endorsing products and services that emphasize agency, self-expression, and social connection – will likely engage the teen consumer's brain effectively and positively.

The Emoji Brain: Short, Symbolic, yet Emotional

The teenage years are a tapestry woven from strands of complex neurological development and succinct social communication. A teenager's aversion to laborious explanations and their preference for brevity in expressing emotions are grounded in brain mechanisms where efficiency is key when processing emotional content. The emotive appeal of simple words like *cool* and *cringe,* or the universal understanding of an eye roll, resonate with the adolescent's neurological predisposition to utilize neural shorthands for emotional expressions.

Researchers find that distinct areas of the brain are especially active during adolescence, such as the amygdala, which processes emotions, and the prefrontal cortex, which manages judgment and self-control. This unique development trajectory explains the teen's susceptibility to images and words that evoke strong emotion – quite literally, these cues often bypass their rational brain and go straight to the emotional one.

This neural predilection is reflected in expressive forms like emojis, which are akin to a modern-day digital lexicon for teens. A multitude of studies suggest that the use of visual media for communication – including emojis and GIFs – is not just a cultural trend but intertwines with the brain's wiring, which favors rapid emotional processing. Marketers attuned to this can use emojis strategically to resonate with a teen audience, creating a brand language that speaks directly to their neurological and cultural compass.

The burgeoning field of computational linguistics shines a light on the impact of these micro-communications. Algorithms assessing sentiment analysis now consider that a single emoji can carry an entire spectrum of emotional subtlety. A well-chosen emoji can encapsulate and convey complex sentiments efficiently, aligning with the teen brain's desire for streamlined communication.

But to be effective, the application of this neuro-shorthand in marketing must avoid condescension or missteps in translation. There's a fine line between joining a conversation and mocking the dialect. Marketers, like linguists deciphering a new language, must immerse themselves authentically into the world of teen emojispeak.

The emoji's directness and universality are its strengths when working within the constraints of the teen neurological toolkit. For example, the simplicity of a thumbs-up or a heart emoji might manifest as intrinsic rewards in the teenager's nucleus accumbens, the same center that responds positively to social validation and inclusion.

There is an elegance in this brevity, one which is deeply entwined with the dopaminergic reward pathways and their connection with the satisfaction of instantaneity.

Moving forward into the kaleidoscope of teen-driven markets, the trend toward visual and expressive economy is not just a fad but an evolutionary step in communication strategies, highlighted by an equally fast-evolving digital language center in the brain. Marketers looking for engagement must respect this new order of communication, crafting messages that honor the economy of emotions valued by the young consumer's active neural networks.

A linguistic revolution is underway, and the teenage brain is its vanguard. The marketer's charge, then, is to be light-footed, quick-witted, and emotionally in tune with the swift currents of teen speak. It takes savvy to stoke a brand with the authenticity that speaks "cool" in the neural circuits of the teen world without trying too hard – remember, overplayed zeal can be as embarrassing as a dad at prom.

In the theater of the teenage brain, where neural shorthand reigns supreme, the effective GenAi algorithms must utilize emojis that are attuned to the quicksilver of teenage emotions. Indeed, there is an open field of emoji creation, and avatar creation is most suited for GenAi and teen engagement.

Pruning – Constantly Changing Networks of Neurons and Friends

During adolescence, a whirlwind of neural pruning occurs throughout the brain, an intricate biological process wherein synapses – the connections between neurons – are systematically strengthened or eliminated based on their utility. In this phase of synaptic "spring cleaning," the brain is honing its neural pathways to improve efficiency and adaptability, employing a "use it or lose it" principle that sees fewer active connections wither away.

This neurobiological undercurrent has a direct impact on teen behavior. As adolescents find themselves prioritizing certain skills and social bonds over others, their brains respond in kind, consolidating neural pathways associated with those frequent activities and interactions. Conversely, the neural groundwork corresponding to neglected experiences faces the synaptic chopping block, highlighting the transient yet critical nature of this developmental period.

Understanding that the pruning process markedly influences behavior helps explain the oft-noted volatility in teenage friendships. The social whirlpool of adolescence isn't merely due to fluctuating hormones or external social pressures; it echoes the neural sculpting that continuously transpires within the teen brain. As the brain refines its vast network, so too do adolescents adjust their social circles, subconsciously mimicking the intricate activity of neural selection happening invisibly within them.

The social tapestry that teenagers weave – expansive at one moment, tightly knit the next – mirrors the ongoing synaptic refinement. Their friendships expand as they explore and strengthen new social connections; they contract as they let go of less fulfilling relationships, paralleling the synaptic elimination of unused neuronal pathways. This analogy frames the seemingly mercurial world of teen social interactions in a new light, providing a neuroscientific backdrop to the shifting landscape of adolescent relationships.

Now, as we pivot from the science to its societal reflection, it is key to consider how this knowledge influences our approach to adolescent development from a practical standpoint. In educational settings, fostering diverse social environments can encourage teens to expand their social repertoire, analogous to training a muscle group to maintain and develop synaptic connections. This is a prime example of how a deeper understanding of neural pruning can guide strategies to enrich the teenage social experience.

From a GenAi viewpoint, grasping these explosive synaptic changes can inspire the creation of influential social campaigns. Algorithms that cater to teenage consumers can cultivate products that emphasize social bonding and self-discovery. Think of a social media platform as a neural landscape: each "like," share, or new friend could represent the strengthening of a synaptic connection. A clever GenAi algorithmic suite plays into this dynamic, designing experiences ripe for sharing and connection, perfectly synched with the synaptic symphony of the adolescent mind.

With a head full of synaptic shears, teens are cutting, shaping, and restyling their social selves daily. It's hard to keep up with the trends – by the time you've mastered *rizz* or *suss*, they're already on to the next phrase. Marketers, akin to savvy shepherds, must lead gently without getting tangled in the ever-shifting yarn of adolescent whims.

Consider this: A keen marketer might craft a campaign that's equivalent to the popular lunch table – appealing, dynamic, and socially rewarding. Here's a free bit of advice – don't be the try-hard brand that arrives late to the meme party wearing last year's slang. Like a perfectly timed high-five in the hallway, marketing to teens must seamlessly blend into their fluid social nexus. It's about being the neural connection that doesn't get pruned, the brand that sticks.

Speculatively, if teen friendships were a stock market, it would be a paradise for day traders and neuroscientists! The hustle and bustle of friendships forming and fading align with the frenetic energy of neuronal trimming within the adolescent brain (Siegel 2015). So, if you're planning to enter this fray with a product or idea, it's essential to tickle the fancy of the teen cortex – be the friend who's invited to all the parties because, let's face it, no one likes last man standing.

In brief, as we tip our hats to the intricate neurology of our teenage kin, we also acknowledge the remarkable adaptability they embody. Through the tempest of synaptic pruning, they emerge social savants, forever reshaping their connections like artists sculpting clay. For those who wish to understand, let alone capture, the teenage zeitgeist, it pays to remember that the art is as transient as the mind it mirrors – dynamically pruning, persistently creating.

Rush of Firsts – Novel All the Time

During adolescence, the human brain undergoes monumental changes that shape cognitive and emotional development. This is a period marked by a whirlwind of neural activity, as synapses strengthen or are pruned in response to experience, sculpting the brain's architecture. One of the most salient changes occurs within the limbic system, specifically the nucleus accumbens, which is heavily involved in the brain's reward circuits and pleasure pathways. These neural pathways are especially activated by new stimuli and experiences, underpinning the teenage propensity to seek out novelty and sensation.

This relentless pursuit of the new – the *firsts* – is emblematic of the teenage experience. From the first loves to the first rebellions, the adolescent brain is deeply influenced by the rush associated with these novel milestones. Dopamine, a key neurotransmitter involved in reward processing, flows liberally within the teenage brain during such experiences, reinforcing the desire to be pioneers in their social circles and beyond.

The pattern of behavior in seeking to be the first, to discover and share, extends to brand interactions and media consumption. When a teenager is the first in their social group to latch onto a new trend or product, it triggers a reward response like other forms of pioneering behavior. For marketers, understanding and leveraging this neural wiring means creating campaigns that position products as gateways to unique and first-time experiences that captivate the teenage need to lead the pack.

Not only is the "what" of these experiences important, but so too is the "how" they are delivered. Musically encoded messages, for example, can create strong associations and recall, given music's profound linkage with memory – especially during formative years. A well-crafted jingle does not just convey a message; it embeds it into the teenage psyche, intertwining with their ongoing neural development.

The heightened emotional vibrancy of adolescence, coupled with a blossoming social identity, means that teenagers are particularly responsive to marketing that matches their emotional frequency. Brands can resonate on a deeper level by aligning their messages with the poignant coming-of-age moments that are common in all teenage individuals' lives – the ones that define the sweet but sharp pangs and ecstasies of growing up.

Therefore, GenAi algorithms must build "first" narratives into product launches or campaigns, encouraging teens to be part of a shared inaugural moment. By creating experiences around these premieres, teenagers are not just purchasing a product – they're investing in social capital, becoming trendsetters among their peers. This capital, in turn, weaves into their evolving personal narrative and identity, a teen's internal story that gains definition with each new first.

Transitioning from science to application, let the symmetry between a teenager's love for innovation and your product reveal itself in an alchemy of firsts. Market the product not just as new but as a milestone – an enviable new dance move, the first footprint on a digital beachhead, or the opening stanza in a modern ode of adolescence. It's about being fresh and lovable like the summer anthem that begs to be put on replay.

In delivering this message, humor and light-heartedness can go a long way. Remember, the adolescent brain is wired to take pleasure in fun and amusement as much as in discovery. A dash of clever humor, couched in the freshness of a first, is a gateway to a teen's good graces. It's about striking that perfect chord – say, launching the first-ever edible headphones, because who hasn't thought their beats could taste as good as they sound?

In conclusion, understanding the neurological underpinnings of the teenage quest for novel experiences provides a unique opportunity to craft marketing strategies that resonate on a deep and impactful level. By aligning your brand with the intrinsic drive for firsts, you forge more than customer loyalty – you weave into the very narrative of their youth, becoming an inseparable thread in the vibrant tapestry of their memories and stories.

Congratulations. You now possess a rare degree of knowledge. You have learned the intricacies and the challenges and the best paths to market to the teenage cohort.

From music to emojis, taste and touch, scent and social media, and much more, you are better equipped than ever to connect with that audience in ways previously unknown.

You now know how to create prompts that wrap advanced neuroscience together with GenAi, to make the most of this unique and powerful technology.

And you can apply that knowledge immediately, for maximum effect.

CHAPTER 4

Gender and the Brain

I n this section we lay out the principles of neuroscience that apply in particular to cisgender women and men.

For purposes of clarity, *cis* refers to biological gender at birth being identical to gender identity in later years.

In this chapter we talk about girls at birth identifying as women in their later years, and boys at birth identifying as men in their older years.

Not enough research has been done on transgender and gender fluid brains, and we truly believe more time and resources to be devoted to understanding this important segment of humanity. For the benefit of humanity, we strongly encourage this investment and research.

With humility, we assert that all learnings in this chapter are just restricted to the section of humanity that is cisgender female and cisgender male alone. For purposes of brevity, all references to female or women, or male or men, in this chapter merely refer to cisgender female and male.

The Cisgender Female Brain

With few exceptions, in country after country women have emerged as perhaps the most powerful and influential consumers of all time.

But modern neuroscience teaches a critical lesson: Reaching and appealing to the female brain requires a deep understanding of the ways in which they are unique, from structural differences to specific functions.

Today though, to achieve the greatest success with your marketing campaigns to women, you must also master the newest science: the dynamic advent of GenAi. It is impossible to overstate the portent and the power of this technology. It is changing everything and so it is important that algorithms understand the nuances of appealing to the female brain. If adequate care is not taken, preexisting societal biases may find their way into GenAi, and we might have cliched approaches to how to make things appeal to the female brain – "Pink it and Shrink it," as it is well documented in marketing. In addition, as the vast corpora that GenAi is trained on is unknown, and mostly undocumented, it is vitally important to explicitly teach algorithms the principles of appealing to the female brain. These teaching form part of guardrails and guiding principles to guide algorithmic output.

In this chapter you will learn chapter and verse about the female brain. You will also learn the ultimate avenue to informing and persuading that brain: the unique combination of GenAi with advanced neuroscience.

That marriage isn't just made in heaven; it is specifically designed to deliver unprecedented results for marketers worldwide.

Your invitation to the female brain awaits. Begin this chapter to accept.

When Emotions Echo: Tailoring Messages to the Female Brain

In the female brain, the insula and hippocampus regions play significant roles, as they are more developed and influential in processing emotional memories (Cahill 2006). The insula integrates somatic and emotional information, contributing to the subjective experience of emotions. Meanwhile, the hippocampus is essential for consolidating information from short-term to long-term memory, often influenced by emotional aspects (Phelps 2004).

Recent neuroimaging studies have revealed that women exhibit greater activation in the insula during tasks involving emotional processing, which suggests a deeper experience of emotions (Domes et al. 2010). Additionally, the hippocampus is involved not only in memory consolidation but also in associating emotional context with memories, resulting in more vivid and enduring emotional memories for women (Andreano & Cahill 2009). By infusing facts and events with emotions, messages become more memorable and impactful due to the enhanced activation of the insula and hippocampus. Therefore, weaving emotional elements into communication is crucial (McGaugh 2003).

Marketing communication tailored for women should tap into the rich reservoir of emotional memory. Therefore, product attributes should be intertwined with stories that evoke feelings, as this can create a vibrant and lasting impression. When promoting a new skincare line, for example, combining factual information with stories of confidence and self-care ensures that the message not only gets heard but also resonates.

Furthermore, it is not sufficient to solely present an emotionally charged narrative; even calls to action should elicit emotions. Emotions have a significant influence on decision-making processes (Bechara & Damasio 2005). Thus, whether it's a prompt to "Join the movement" or "Experience the transformation," coupling the action with an emotional component generates a more powerful and personally aligned response.

Conversing with Compassion: The Collaborative Cadence of Communication in Female Brains

Communication has always played a vital role in society, from the dawn of humanity to the bustling digital age. However, beneath the surface, a nuanced distinction arises from the very foundation. Through studies, it has been revealed that girl babies exhibit a threefold increase in turn-taking during play compared to their male counterparts, highlighting an inherent collaborative spirit woven into the architecture of the female brain (Geary 1998).

The prefrontal cortex, crucial for complex behavior and social interaction, exhibits structural and functional differences that predispose females toward collaboration (Blakemore et al. 2010). The female brain, with its inclination for collaboration, not only attends to the content of speech but also the manner in which it is delivered. Language transcends mere information exchange, transforming into a ballet of shared intentions and communal goals (Maccoby 1998).

In the female brain, the mirror neuron system is particularly pronounced and correlates with empathic accuracy, allowing for a deep mirroring of others' states and facilitating compassionate communication (Kaplan & Iacoboni 2006).

Engaging the female audience goes beyond simply presenting a product or slogan; it involves inviting a dialogue and fostering participation. Rather than issuing commands, marketers can pose questions that encourage collaboration. The modus operandi becomes "Shall we explore this together?" instilling a narrative of unity and camaraderie. This creates a space where mutual decision-making flourishes in resonance with the collaborative frequencies of the female brain (Gallucci & Perugini 2003). Migrating from directional to collaborative communication is appealing to the female brain.

Authentic engagement is key to effectively harnessing the power of collaborative language. Superficial interactions fall short of creating the same harmonious chords as personally appreciating someone's unique contribution. In communication, the illusion of togetherness is not enough – it is the heart-to-heart connection that captivates (Cross & Madson 1997).

The Eyes Have It: Unlocking the Visual Secrets of the Female Brain

The significance of eye contact in human interaction has long been acknowledged. However, a deeper understanding of the cognitive neuroscience behind gaze perception reveals intricate complexities in how females process eye contact. Studies have shown that the eye contact effect, closely tied to attention, attunement, and social awareness, may have distinct nuances in the female brain.

The exploration of gaze perception begins in the brain's superior temporal sulcus (STS), a region responsible for discerning the direction of others' gaze. This neurological process is crucial for social cognition as it enables individuals to understand focus and intentions. For females, who are more attuned to nonverbal cues, deciphering gaze direction goes beyond visual perception; it serves as a gateway to empathy and connection.

The fusiform gyrus – an area involved in facial recognition – exhibits notable activation when processing direct eye contact compared to averted gazes. Females, in particular, demonstrate an amplified neural response to eye contact, aligning with their nurturing and kinship-oriented social roles. This inherent response holds significant implications for emotionally driven engagements and marketing techniques.

Females are more likely to be captivated by advertisements that employ direct gaze, activating the same neural pathways involved in real-life social interactions.

By leveraging the instinctual pull of the eyes, marketers can establish an invisible connection that grabs the shopper's attention more effectively than any compelling tagline.

Applying this scientific knowledge to marketing strategies suggests that incorporating images featuring direct eye contact with the viewer increases the likelihood of capturing their attention – akin to a "visual handshake." Through the engagement of direct gaze, the observer becomes an active participant in a two-way dialogue, receiving a personal invitation into the narrative of the advertisement or brand.

However, effective eye contact in marketing goes beyond a mere staring contest. Like in any meaningful conversation, there needs to be depth and expressiveness behind those eyes. The emotion and expressivity conveyed through a model's gaze can tell stories, evoke sentiment, and exude charisma without uttering a single word. It becomes a silent form of communication that resonates deeply with the shopper's emotions.

Mirror, Mirror: The Reflective Magic of Female Neural Circuitry in Marketing

Similar to the mystical Mirror of Galadriel in Tolkien's Middle Earth (Tolkien 1986), which reveals distant truths and future possibilities, the "mirror neuron system" in the human brain reflects the actions and emotions of others. This network serves as the neurological basis for imitation and empathy. Surprisingly, recent studies indicate that the mirror neuron system is more pronounced in women, endowing them with an enhanced ability to empathize and imitate. This heightened mirror neuron activity in females may be evolutionarily linked to nurturing roles and social bonding.

Examining this reflective neural phenomenon, scientists have discovered that mirror neurons activate not only when a person performs an action but also when they observe someone else engaged in the same action. For women, this mirroring experience is akin to a neural dance. They are not merely watching someone enjoy a delightful meal or cradle a new product; they are vicariously experiencing it, feeling the textures, tastes, and emotions as if they were their own.

The parallel evolution of social behavior and brain physiology reveals the vital role of physical proximity and touch in fostering mutual understanding and connection. Neuroendocrine systems involved in processing physical closeness, such as oxytocin, actively contribute to the formation of social bonds and maternal behavior. This chemical catalyst for connection, often highlighted in mother-infant relationships, also operates in the everyday interactions between women and their social groups.

Marketers can leverage the activation of mirror neurons in women through visuals that depict shared enjoyment and closeness, incorporating proximity and touch. By creating content that showcases women engaging in acts of unity and pleasure, marketers can create a neural resonance with viewers. The potential customer is drawn in by the simulated experience, prompting her mirror neurons to seek a sense of kinship and merriment, fostering a perceived connection with the product.

Applying this knowledge to advertising practices involves capitalizing on humanity's innate tendency to mimic others. When targeting female demographics, it is strategic to move away from imagery depicting isolation or individual indulgence. Instead, advertisements can flourish by featuring groups of friends sharing moments, where the fondness and companionship overflow. These mirror neuron interactions need to be triggered through GenAi algorithmic creations both in prose and imagery.

The Neuro Dance: Decoding the Cortisol Waltz in the Female Brain

The hormone cortisol, commonly known as the *stress hormone*, circulates through our bloodstream in response to stressors, directed by the intricacies of the hypothalamic-pituitary-adrenal (HPA) axis. While both men and women experience cortisol's effects, the female brain exhibits a unique sensitivity and response to this hormone. This distinction is rooted in biology, as estrogen, the primary female sex hormone, influences the HPA axis, resulting in a different interplay between stress and the endocrine system in women. Elevated levels of cortisol can have a negative impact on mood and behavior.

In examining the stress response further, the amygdala emerges as a key player in the emotional dynamics of the brain. In women, the amygdala's response to negative stimuli is often amplified, making stress-induced emotional choreographies particularly profound. The interaction between cortisol and the amygdala resembles a pas de deux, leading to heightened emotional consequences following stressful events in women.

Neuroimaging studies have revealed increased connectivity between the amygdala and other regions involved in emotion processing in females. This enhanced connectivity suggests that cortisol conducts a more intense emotional experience, pulling emotional chords strongly in response to stress within the symphony of the female brain.

For marketers, these insights have clear implications. While advertisements typically follow a stress-resolution narrative arc, it is important to recognize that the female audience may experience an unpleasant intermission during the stressful sections. Hence, the storyline of advertisements needs to be rewritten, incorporating subtler depictions of stress and more nuanced tensions to prevent cortisol levels from spiraling out of control.

In the realm of advertising, it becomes crucial to modulate the drama. Instead of employing shock tactics that induce cortisol spikes, marketers must choreograph a ballet that balances tension with tranquility. Campaigns that navigate away from stressful territory and embrace more positive storylines are more likely to make a lasting and positive impression on female audiences. As GenAi increasingly creates marketing materials, advertising copy, and storylines intended to appeal to female audiences, cortisol moments must be carefully orchestrated to continually engage and enthrall the female brain.

The Many Hats of Hera: Unraveling Female Multitasking Mastery

The ability to multitask is often associated with a mental juggling act, but neuroscience reveals that this skill is deeply rooted in cognitive architecture, particularly in females. Research suggests that women exhibit superior multitasking capabilities compared to men, pointing to biological and cognitive frameworks that support this skill (Mäntylä 2013).

One potential explanation for this proficiency lies in the distinct neural strategies employed by females when handling simultaneous tasks. The corpus callosum, a bridge that facilitates communication between the brain's hemispheres, is proportionally larger and denser in females. This structural advantage provides a more efficient pathway for information transfer, potentially facilitating smoother coordination of dual tasks and enhancing multitasking abilities (Gur et al.1999).

The prefrontal cortex (PFC), often referred to as the brain's executive suite, plays a vital role in planning and orchestrating thoughts and actions based on internal goals. In women, the PFC shows greater activation during multitasking, suggesting a heightened level of organization and control when managing multiple threads of thought and action (Szameitat et al. 2015).

Examining the biochemical aspects of multitasking, estrogen receptors are found in key regions involved in this skill, including the PFC. Estrogen, a hormone more abundant in women, interacts with these receptors to modulate cognitive function. It has been linked to improved working memory and cognitive flexibility, key components of effective multitasking. This insight into the neural underpinnings provides a glimpse into why women may excel in multitasking (Hampson 1990).

From an evolutionary perspective, women may have developed superior multitasking skills due to the demands of managing offspring, foraging for food, and maintaining complex social networks throughout history. The ability to multitask efficiently would have been advantageous for the survival of tribes, leading to the development of this skill.

Depicting women seamlessly handling multiple tasks resonates with their experiences. Showcasing a woman effortlessly transitioning from a business call to meal preparation to overseeing homework aligns with the audience's reality and cognitive predispositions.

Instead of presenting multitasking as a chaotic circus act, a lighthearted approach can emphasize the normalcy and grace with which women manage multiple tasks. Such advertisements tap into viewers' cognitive empathy by recognizing familiar patterns that their own brains navigate on a daily basis.

Connecting the Dots: The Female Brain and the Long Game in Marketing

In the field of cognitive neuroscience, there is a prevailing belief that the female brain possesses a natural aptitude for "big picture" thinking. This skill allows individuals to construct and comprehend complex scenarios and foresee how present actions connect to future outcomes. While this neural ability is not exclusively gender-specific, studies have indicated that females are more inclined toward long-term planning and considering the broader implications of their actions (Andreano & Cahill 2009). This inclination is rooted in neurological patterns that stimulate visionary thinking and multifaceted cognitive processing.

A critical element of this neural proclivity resides in the prefrontal cortex (PFC), which serves as the brain's hub for executive functions, such as planning, decision-making, and regulating social behavior (Miller & Cohen 2001). Female brains often utilize the PFC to seamlessly integrate various pieces of information and mold them into a coherent narrative that extends into the future. Additionally, the PFC collaborates with the limbic system to imbue these narratives with emotional significance, highlighting not only the passage of time but also the resonance of experiences throughout that journey.

From an evolutionary perspective, it is hypothesized that this long-term perspective in females developed as an adaptive response to the demands of nurturing offspring and maintaining social cohesion within complex community networks (Geary 1998). By integrating immediate needs with long-term goals, females were better equipped to ensure the well-being and survival of both their children and their community.

Estrogen receptors, densely populated in key cognitive areas like the PFC and hippocampus, enhance cognitive flexibility and the ability to strategize (Hampson 1990). These receptors respond to fluctuating hormone levels and influence how women perceive and contextualize events within a larger framework, reinforcing their inclination toward comprehensive chronological mapping.

This scientific backdrop provides insights into how female consumers approach decision-making and preference. When presented with a product, their consideration extends beyond the object itself, delving into the narrative space where the product's place in their lives unfolds over time. Consequently, marketing that solely emphasizes immediate gratification may not resonate as strongly with women, who seek connections that integrate their purchases into the tapestry of their ongoing stories.

To leverage these insights, marketers should develop campaigns that construct a storyline around their products. The purchase should not be portrayed as the culmination but rather a pivotal chapter in a longer epic. For example, a narrative positioning a particular brand as not just a breakfast option but as a companion that energizes mornings to fuel ambitions and achievements would resonate with the future-oriented thinking of the female brain.

Queries over Commands: The Female Brain's Linguistic Harmony

From a visual marketing standpoint, imagery should suggest continuity, such as a literal roadmap where today's savored beverage or cherished fashion item becomes the starting point of a path leading toward a sunrise of possibilities. This approach aligns with the ideation patterns in female neural circuitry, validating present moments and elevating them as sequences in a larger journey.

Kuno and Kaburaki's research (1977) on communication pragmatics highlights the significance of shared knowledge and cooperative principles in driving language use. Women tend to employ "conversational maintenance work" more frequently, guiding interactions with questions to maintain rapport and elevate the listener's role from passive recipient to active participant. This behavior is supported by enhanced neural circuitry involved in social cognition and empathy, as evidenced in studies on gender differences in the brain (Baron-Cohen et al. 2005).

The anterior cingulate cortex (ACC) and orbitofrontal cortex (OFC), which play roles in decision-making and social interaction, exhibit structural and functional variations between genders. In females, these regions display heightened activity when engaged in tasks requiring collaboration and understanding of others' perspectives, aligning with the behavioral inclination toward cooperative language. These neural networks make a questioning style more harmonically resonant with the cognitive architecture of females.

When applying this scientific understanding to the realm of marketing, it becomes evident that a "soft touch" is necessary in communication strategies targeting women. Instead of employing commanding imperatives, crafting promotional materials with gentle leading questions that resonate with the latent desires and needs of the audience becomes a more effective tool. This approach aligns with neuroscientific findings and harmonizes with the relational and interconnected worldview held by many women.

Marketing should adopt a tone that feels like a shared conversation rather than a loud announcement. This conversational dance extends from the supermarket aisle to the family dinner table, where choices are contemplated, and products become partners in the choreography of daily life. Marketing becomes a dialogue, an exploration of wants and wonders, inquiring, "What's your pleasure?" rather than proclaiming, "Here's what you need."

The Emotional Palette: Decoding Subtlety in the Female Mind

The female brain possesses an exquisite ability to perceive emotions with precision, akin to discerning the softest whispers in a gusty breeze. Key players in this emotional acuity are the amygdala, responsible for decoding emotional cues, and the insula, which facilitates empathic resonance (Adolphs et al. 1995; Singer et al. 2009).

The amygdala expertly interprets emotions from facial expressions and body language, often working beneath our conscious awareness. Women have been found to have superior emotional recognition capabilities compared to men in numerous studies, suggesting a heightened sensitivity to emotional signals (McClure 2000).

Moving through the cerebral canvas, the insula brings subjective emotional experiences into sharp relief. With its extensive neural connections in females, the insula fosters a deep understanding of others' feelings, allowing women to navigate social dynamics with intuitive awareness of hidden emotional currents (Craig 2009).

The synergistic relationship between the amygdala and insula grants women an enhanced emotional literacy, enabling them to detect sincerity, sarcasm, or sadness in subtle nonverbal cues. Hormonal influences, such as estrogen, further fine-tune the female brain's response to emotional stimuli, emphasizing their inclination toward emotional depth and complexity (Derntl et al. 2008).

In the realm of marketing, understanding and acknowledging this sensitive emotion-driven compass can be the foundation of impactful messaging. Products and campaigns that evoke genuine emotions can deeply engage women by speaking the nuanced language their brains are wired to understand. A heartfelt commercial that elicits laughter or tears can create lasting emotional memories associated with the brand.

The Cisgender Male Brain

The male brain is a complex and fascinating entity, with its intricate neural networks influencing decision-making, behavior, and cognitive processes. In this chapter, we delve into the neurological foundations of the male brain and its implications for marketing strategies. From understanding the interplay of emotional and logical tendencies to navigating visual-spatial processing, we explore how insights from neuroscience can shape effective marketing targeted at the male demographic.

The insights are divided into five sections, each focusing on a specific aspect of the male brain and its relevance to marketing. We begin by examining the neural mechanisms that guide decision-making and emotional responses, shedding light on how these insights can inform marketing narratives. Earning a deeper understanding of the male brain's systemizing tendencies and emotional responses provide valuable insights for marketers seeking to connect with male consumers.

We then explore the nuanced role of the male brain in competitive scenarios and the interplay between defense and dominance, offering insights into how these tendencies can inform persuasive marketing strategies. The following section delves into the male brain's neurobiology of worry, highlighting the role of the anterior cingulate cortex and its implications for communication and marketing messaging. This is followed by an exploration of the male brain's visual-spatial processing capabilities, providing insights into how marketers can leverage these skills in engaging male consumers. We round off the chapter by unraveling the neural predispositions of the male brain in relation to impulsivity, memory, and visual-spatial processing, and how these factors can shape effective marketing techniques.

Each section offers a blend of scientific understanding and practical applications, aimed at providing marketers with a nuanced and informed approach to targeting the male demographic.

Bridging the Gap: Navigating Emotion and Logic in the Male Mind

It is often presented as a simplistic dichotomy – the logical male brain versus the emotionally attuned female brain. However, neuroscience reveals a more nuanced picture that challenges these broad generalizations. While there are differences in brain architecture between males and females, the distinctions are not as clear-cut as popular culture suggests. The male brain does show a propensity for systemizing tendencies, such as a preference for patterns, analytics, and logical sequencing. When faced with emotional issues, these inherent biases can lead to a focus on problem-solving approaches.

It is important, however, not to misinterpret this systemizing trait. Emotional processing does occur in the male brain, although it often emphasizes logical resolution. The prefrontal cortex (PFC), as previously mentioned, plays a crucial role in decision-making processes and also acts as an intermediary between emotional impulses and rational deliberation. When confronted with an emotional dilemma, men may approach it from a resolution-oriented perspective rather than an empathetic one, explaining the tendency to try and "fix" rather than empathize (Deng et al. 2016).

However, emotional states can significantly influence cognitive processes in men. Understanding how this interplay between emotional and logical tendencies can impact effective communication is essential, both on a personal and marketing level.

Bridging the gap between emotion and logic requires a delicate dance, particularly in the realm of marketing. When engaging male audiences, a marketer might choose to highlight the logical features of a product while simultaneously tapping into the emotive reasons that make it desirable. Crafting a narrative that follows the hero's journey archetype, blended seamlessly with clear information, appeals to both cognitive efficiency and emotive engagement (Alexander 2017).

In practical terms, this means GenAi must focus on creating advertisements that showcase appealing narratives aligned with aspirations, success, or protection, while seamlessly weaving in technical specifications. This approach leverages both the analytical brain, satisfied by the product's features and benefits, and the emotional brain that identifies with the thematic underpinnings of the story. It exemplifies the fusion of data and drama.

The power of demonstration should not be underestimated either. The mirror neuron system suggests that observing actions activates corresponding neural pathways in the observer. Showcasing relatable characters using the product not only informs but also creates a vicarious experience (Lacoste-Badie et al. 2014). In the male brain, this primes the mental mechanics of action, aligning the observed utility with personal applicability.

A campaign that combines observational learning with structured information embraces both the logical predispositions and empathetic capabilities of the male brain. For example, a car commercial that highlights not only the vehicle's performance but also the life experiences enabled by it strikes chords on both ends of the cerebral spectrum.

The Neuronal Knight: Defending the Realm of the Male Mind
The male brain, characterized by a larger dorsolateral prefrontal cortex (DPN), has been linked to heightened readiness for defense, fear, and aggression. This inclination influences behavior across various contexts, from personal confrontations to competitive sports.

Hormones, particularly testosterone, play a significant role in modulating these tendencies. Testosterone levels can affect the amygdala, a region associated with fear and aggression, and subsequently influence a man's demeanor and decision-making in the presence of perceived threats or competition.

Understanding the neural basis of these tendencies provides a foundation for creating compelling marketing strategies. GenAi algorithms can leverage the heightened response to competition by incorporating sporting metaphors revolving around victory and honor, concepts that deeply resonate with the male psyche and evoke feelings of competitiveness and camaraderie.

Sports serve as a powerful allegory for business and life, allowing marketers to connect with the male audience on a profound level. Car commercials, for example, often depict vehicles overcoming rugged terrain or outperforming competitors, symbolizing strength, control, and dominance, which appeal to the male sense of mastery and adventure.

However, this approach must strike a nuanced balance between aggression and protection, the visceral and the virtuous. Products and services can be positioned not only as victories achieved but also as shields guarding against threats, whether financial security, health, or family well-being.

Campaigns that solely emphasize aggression and defense can become tiresome. Skillful marketers infuse humor into the narrative, providing a respite from life's battles. Lighthearted quips about "fighting" mundane tasks or "defending" against the dullness of everyday products bring a friendly and relatable tone to campaigns, enhancing memorability and approachability.

When GenAi narratives portray men triumphing over adversity or protecting their loved ones, mirror neurons are activated, causing viewers to vicariously experience the feelings of accomplishment and security. This phenomenon influences product appeal and brand loyalty, as seeing becomes a form of feeling and doing.

GenAi messaging that appeals to men requires bridging the worlds of science and storytelling. By aligning with men's instinct to compete and protect, narratives should resonate on both a visceral and intellectual level. Addressing these intrinsic drives encourages male audiences to perceive products as tools that empower them to overcome challenges and safeguard what is important to them.

Appealing to men in the marketplace means embracing their chivalrous nature within the realm of the cerebral. By tapping into the male brain's duality of defense and dominion, marketers can create campaigns that are not only compelling and convincing but also endearing. In the world of marketing, when done right, it isn't an assault on the senses but a call to adventure, where products become both the sword and the shield in the consumer's hand.

The secret to breaking through in the male market is found at the intersection of science and narrative. Marketing becomes an homage to the ageless legends of heroism that continue to enthrall modern man by adopting the stance of respecting the twin urge for protection and control. It's not a ploy; rather, it's an invitation to take a fascinating trip where products turn into necessary resources for guaranteeing security and overcoming obstacles.

Warrior Mindset: Marketing to the Male Sense of Self
Testosterone, commonly known as the "male hormone," exerts a significant influence on the brain, shaping perceptions and actions with a potentially more combative or competitive edge. This hormone interacts with various brain regions, including

the amygdala, to modulate social behavior, aggression, and the navigation of hierarchies and challenges. Such interactions have been observed across different animal species and human social structures (Geniole & Carré 2018).

In marketing, it is important to engage these aspects of the male brain without solely appealing to base instincts. While the competitive drive in men can serve as a starting point for messaging, it is essential not to rely solely on provoking aggression. A refined approach would involve stimulating the motivation to achieve and prevail, offering triumph without invoking conflict. This could include aspirations toward success within a community, personal development, or mastery.

Advertising narratives that resonate with men often portray a journey of overcoming obstacles to achieve a desired goal, aligning with the timeless "hero's journey" archetype. Such narratives tap into deep-rooted psychological patterns that transcend gender and are particularly powerful in male storytelling preferences (Peterson et al. 2017). Stories that highlight triumph not only entertain but also engage listeners on a primal level, activating the brain's pattern recognition systems.

When applied to marketing, these triumph-centric stories do more than entertain; they engage male consumers. By framing their product or service as a catalyst for unlocking potential, solving challenges, or contributing to a journey toward victory, marketers can motivate men. The narrative should present the protagonist not just using the product but showcasing how it enhances their ability to strive and thrive.

It is important to note that men, like all humans, are driven not only by competition but also by cooperation and the bonds formed through shared endeavors. Research in psychology has explored the cooperative aspect of the male psyche, highlighting how camaraderie and teamwork activate reward pathways similar to those associated with personal achievements. Marketing that captures the spirit of teamwork and communal success, whether through sports analogies or narratives of shared challenges, can be highly effective.

Clever, challenging, and surprising humor activates the brain's reward circuitry. This not only leaves a lasting impression but also balances the intensity of competitive narratives with a lighthearted touch. Incorporating clever banter, puns related to product utility, or amusing scenarios can invigorate marketing campaigns and connect with men on a different, yet potent, cerebral level.

Visual storytelling that mirrors the actions of protagonists can also have a profound impact on men. Seeing others engage in behaviors or use products of interest activates the brain's mirror neurons. These neurological mimics are part of the mechanism behind the persuasive power of demonstrations, particularly for men who value practical applications of products.

What, Me Worry? Untangling Worry in the Male Mind   The scientific understanding of the anterior cingulate cortex (ACC), located deep within the frontal lobes of the brain, has been evolving. Recent neuroimaging studies reveal the ACC's significant role in cognitive functions like error detection, task anticipation, attention, and decision-making. Additionally, it plays a role in regulating mood, empathy, and the expression of emotions. Considering gender differences in brain function and their impact on behavior and propensity

for anxiety adds an intriguing layer to our understanding of the nuanced role of the ACC.

The idea that a smaller ACC in males leads to less worry has been of interest in neuroscience. However, as our understanding of brain-behavior relationships deepens, this notion has become more complex (Bekhbat & Neigh 2018). ACC volume alone does not solely determine one's propensity for worry. Other factors, such as the intricate networking with other brain structures, neurochemical modulation, and environmental influences, also contribute to anxiety-related behaviors. While men may, on average, exhibit less rumination and anxiety, it is essential to recognize the wide spectrum of individual differences. Many men experience significant anxiety, and many women display strong analytical thinking without excessive worry.

Looking at the androgen receptor gene carried by men, studies suggest a potential neural efficiency in pathways involving the ACC. Men's brains often demonstrate greater cognitive task efficiency, utilizing fewer resources. This efficiency may partially explain the observed differences in expressing worry or anxiety. However, it is crucial to acknowledge that individuals vary greatly, and many men do experience anxiety, while numerous women exhibit analytical thinking free from excessive worry.

Translating this scientific knowledge into practical applications highlights the value of understanding the ACC's role in worry, particularly in high-stress decision-making environments like business or emergency services. Leveraging this understanding, training programs could be developed to enhance resilience and cognitive flexibility in such scenarios, benefiting individuals regardless of gender.

The marketing world, with its focus on connecting with male customers, can take these neural insights into account. Rather than relying on fear-based marketing strategies that play into anxieties, narratives emphasizing resolution and action may be more effective. Choosing an ad that portrays proactive solutions instead of inciting dread about life uncertainties may resonate more strongly with a male audience (Putrevu 2001).

It is also important to note that an awareness of these neural tendencies can foster understanding and compassion in interpersonal relationships. Someone with a brain less prone to worry may not be less caring; they simply approach concerns differently. Communication strategies, whether in personal relationships or targeted marketing campaigns, can benefit from emphasizing action and resolution rather than dwelling on potential negative outcomes.

Ultimately, understanding the complexities of the brain, such as the role of the ACC in worrying, requires seeking balance. Recognizing that worry and resolution exist on a spectrum across individuals helps us appreciate the individual differences and navigate the social world with greater empathy. In this equilibrium, it is crucial not to overlook the subtle contributions of the ACC to human behavior – a point marketers and leaders should take note of.

Visual Processing: Pictures Worth a Thousand Words   Understanding the male brain's approach to language and visual processing has practical implications for GenAi algorithm and output design. While it is an

oversimplification to suggest that men inherently prefer visual information over verbal communication, it is true that the human brain, in general, processes images faster than words (Potter et al. 2014). Consequently, imagery in advertising can often be a more immediate and impactful way to convey a message for most people, including men.

Moreover, it is important to challenge the belief that men struggle to process extensive verbal information (James 2015). Neuroscience highlights the brain's preference for efficient cognitive processing, rather than an intrinsic aversion to lengthy discourse, regardless of gender. This efficiency explains why bullet points and infographics are generally more effective than dense paragraphs in advertisements.

It is crucial to update the antiquated view that the male brain is inept at multitasking. Recent research indicates that the differences in multitasking abilities between genders are negligible (Lui et al. 2021). The idea that women excel at multitasking while men cannot handle more than one task at a time is a myth rather than science. Instead, multitasking abilities depend on context and individual capabilities, with some individuals, both men, and women, naturally excelling at multitasking and others finding it less comfortable.

That being said, many men (as well as women) find a focused approach more effective for cognitive performance. Neuroscience has demonstrated that goal-oriented tasks that require sustained attention and direct engagement often lead to better outcomes. This preference for goal-oriented and directed tasks is not exclusive to men but is a general strategy for enhanced performance that applies to marketing campaigns aimed at any audience.

From a GenAi standpoint, it is about finding the right balance between engaging imagery and clear, concise language when targeting men. Coupling bold visuals with focused messaging helps cut through advertising noise, capturing attention without overwhelming cognitive load. However, these elements must be harmonized because a stunning image without a coherent message is as ineffective as great copy lost in a sea of text.

While taking into account these neurological nuances, it is worth addressing the supposed aversion to multitasking among men. The irony of a man arguing vehemently about his single-tasking skills while simultaneously watching sports, flipping through a magazine, and snacking is not lost on marketers. Rather than avoiding multitasking in marketing strategies, showcasing products that fit into a potential buyer's existing rhythm of multiple activities could be a clever approach.

In practice, GenAi narratives that leverage the brain's efficiency will captivate without overwhelming. For men, as for anyone else, a compelling story matters, and making that story visual, straightforward, and goal-oriented may be the key to engagement. A campaign that can convey a journey, a challenge overcome, or a victory achieved in just a few impactful words and a powerful image merges the art of marketing with the science of the male brain.

Impulsivity: Seeds of Recklessness
Dopamine, a neurotransmitter associated with reward and pleasure, plays a significant role in the brain's reward systems. Men's brains may exhibit higher dopamine activity in certain areas, which could contribute to a propensity for risk-taking behaviors (Munro

et al. 2006). However, this tendency is not a universal trait among all men. Impulsivity is influenced by the interplay of neurotransmitter systems, situational factors, and learned behaviors.

Considering these scientific insights, marketers targeting the male audience must challenge long-held myths and stereotypes that have shaped marketing strategies. Instead of assuming men are inherently impulsive and disinterested in details, it is crucial to recognize individual variability and understand that male consumers, like any consumers, are influenced by personal values, cultural norms, and the context of the purchasing environment.

GenAi algorithms can leverage these nuances by creating gamified shopping experiences that appeal to individuals who are more responsive to dopamine. However, caution must be exercised to avoid overgeneralization, as what excites one individual may repel another. Understanding an individual's neurochemistry can guide more personalized marketing approaches.

How many men have bid on an eBay item, not because they needed it, but because the bidding was closing soon. Impulsivity is part of the male brain makeup, and GenAi algorithms need to capture that in their output.

Spatial Processing: Why Men Never Ask for Directions   The neurological aspects of spatial awareness and visual processing in the male brain present intriguing insights. One area of interest is the parietal lobe, specifically the intraparietal sulcus, which plays a crucial role in spatial orientation and navigation. Research suggests that this brain region's proficiency in interpreting spatial information may underlie men's enhanced performance in tasks requiring spatial acuity and mental manipulation of objects (Yuan et al. 2019). Functional magnetic resonance imaging (fMRI) studies have further supported these findings by demonstrating heightened activation of these regions during tasks like virtual navigation and geometric puzzles.

Another significant cognitive aspect is the superior parietal lobule, which is essential for visuo-motor coordination and the integration of sensory information into a coherent representation of space. Men often exhibit more pronounced brain activity in these areas during complex spatial tasks compared to women, potentially explaining their proficiency in visual puzzles and spatial challenges. While individual differences should be acknowledged, these trends shed light on the male bias toward visual and spatial problem-solving.

The preference for visual processing extends to the realm of optical illusions. Differences in neural architecture suggest that the male brain may process visual cues like size, contrast, and perspective more efficiently than the female brain (Spierer et al. 2010). However, these findings must be interpreted with caution, recognizing that social conditioning and experiences also shape perceptual skills. Nevertheless, the structural predispositions of the male brain provide insights into their engagement with visual trickery.

Moving to practical applications, these neural predispositions have real-world implications. In fields like architecture and engineering, where spatial reasoning is paramount, understanding the male brain's receptiveness to spatial configuration can inform educational strategies. Tailoring pedagogical techniques to capitalize on these inherent strengths may optimize learning outcomes. Recognizing these neural tendencies allows for the creation of environments that nurture and challenge the male inclination for spatial tasks.

GenAi algorithms that exploit the male inclination toward visual information can be instrumental. Advertisements utilizing strong visual elements like dynamic graphics or interactive demonstrations can tap into the male brain's natural propensity for processing visual information. Campaigns featuring 3D product models or virtual reality experiences can engage the male demographic, allowing them to establish a spatial relationship with the product. The irony lies in the fact that a man might find more excitement in a 3D rendering of a product than the product itself.

CHAPTER 5

The Maternal Brain

For many women, life both before and after childbirth brings fundamental changes in basic brain structure and chemistry.

Modern neuroscience has delved deeply into those changes – why and how they occur, and their import for every aspect of her life.

In this chapter, we will plunge into the mysteries and magnificence of a mother's mind. And specifically, how marketers can craft products and messaging that appeal most effectively to her. You will learn why humor is so important. Why her basic human orientation toward "fight or flight" morphs into "tend and befriend." Plus how her five senses are altered – permanently. And much more.

The Maternal Instinct: Nurturing the Future

The maternal instinct, combining intuition and a reflex to protect, is rooted in evolved brain structures such as the insula. The insula plays a vital role in emotional processing and empathy, allowing mothers to detect even the most subtle signals of their child's needs and respond with almost precognitive accuracy.

The insula is particularly remarkable in its involvement in interoceptive awareness, which refers to perceiving the body's internal state. This heightened awareness enhances a mother's intuition about her child's well-being (Bjertrup et al. 2019). When her infant cries, a mother displays an uncanny ability to discern whether it is due to hunger, pain, or a need for attention. This level of perception also extends to detecting subtle discrepancies in the environment, serving as an early warning system for potential threats to her children's safety.

Given the well-developed insula of mothers, it is crucial for marketers to communicate with transparency and authenticity. Attempting to mislead or exaggerate may lead to distrust. Therefore, campaigns should be built on clarity and authenticity to ensure that each message aligns with a mother's innate intuitive understanding.

Hormonal changes that accompany motherhood further contribute to the maternal instinct. Oxytocin, often known as the "love hormone," plays a crucial role in establishing and strengthening social bonds, particularly in mothers. It helps initiate the maternal attachment process and increases caregiving behaviors (Kohlhoff et al. 2017). Oxytocin also sharpens a mother's social awareness, making her highly attuned to the social world and motivating her to seek connections with experienced peers or family members who can offer support and guidance.

This scientific understanding explains why social networking platforms hold appeal for mothers as they satisfy their biological desire for community and connectedness. Recognizing this, online forums, brands, and marketers can create spaces that resonate with a mother's strong need for connection and exchange. By facilitating discussions, sharing insights, and fostering a virtual community reminiscent of a modern-day village, marketers can engage mothers more effectively.

GenAi efforts targeting mothers involve not only making them laugh but also creating shared moments of understanding. When a campaign can playfully acknowledge the absurdities of parenting, it not only entertains but also fosters deep connections. This approach transforms a brand from a mere entity selling products into an empathetic companion on the motherhood journey.

The adaptability of the maternal brain is another remarkable aspect. In the digital age, mothers have an array of resources at their disposal, including apps, podcasts, and online platforms. They seamlessly integrate digital assistance into their parenting practices, employing technology to augment their instinctual caring instincts. Mothers rely on technology not only for practical purposes but also for personal growth and problem-solving.

The Mother's Mind: Understanding the Transformation

The experience of becoming a mother not only changes a woman's title but also alters her brain structure. During pregnancy and the postpartum period, a remarkable process called *neurogenesis* takes place, resulting in the generation of new neurons in her brain. This is an extraordinary phenomenon, as the adult brain typically replaces lost neurons rather than creating new ones (Medina et al. 2020). This surge of neural proliferation equips mothers with heightened cognitive capabilities, enabling them to protect and nurture their offspring better.

This blossoming of the brain is not just a marvel of nature but also an opportunity for marketers. It is important to recognize that the neuroplasticity of mothers is not static but continuously evolving, particularly during the journey of motherhood. Whether marketing gadgets, nutritional supplements, or learning tools, it is crucial to understand that a mother's brain is primed for absorption, learning, and adaptation.

The maternal brain is finely wired for building rich social networks, which are crucial for the survival of both the mother and child. Hormonal changes during pregnancy enhance the brain's focus on social cognition, enabling new mothers to identify reliable social partners and discern them from less dependable ones (Zhang et al. 2020). Establishing trust within her social sphere, often referred to as her tribe, is of utmost importance as it provides a foundation for survival during vulnerable periods of child-rearing.

Scientific research reveals a fascinating neurological transformation that occurs when women become mothers. Neuroscientific studies have uncovered that motherhood triggers a remarkable display of hormonal influences and structural changes in the brain. A study conducted by Hoekzema et al. (2017) highlights significant alterations in gray matter volume in regions associated with social cognition and theory of mind, indicating an enhanced ability of mothers to understand the needs and intentions of their children. Such changes also amplify a mother's aptitude for empathy and bonding, facilitating a close and intimate mother-infant relationship. Moreover, memory circuits in the brain undergo rewiring.

Findings from both behavioral and rodent studies suggest that mothers become more efficient at multitasking, exhibit improved working memory, and become more attuned to environmental cues (Rutherford et al. 2018). Therefore, what may be perceived as "mommy forgetfulness" is, in fact, a strategic prioritization driven by the brain's focus on essential maternal responsibilities.

The evolution of the mommy brain extends beyond cognitive enhancements, impacting consumer behavior as well. Research indicates that new mothers develop heightened vigilance toward child safety and well-being. This translates into a consumer persona characterized by selectiveness, a quest for information, and a strong dependence on networks. Product endorsements, safety information, and recommendations hold immense value and influence for these mothers. To effectively market to the mommy brain, it is crucial for marketers to recognize that mothers are not merely seeking products but stories that resonate with their emotionally enriched and cognitively transformed experiences. Functional attributes must intertwine seamlessly with narratives that evoke feelings of security, trust, and nurturing. Additionally, incorporating humor into marketing strategies can tap into the comical aspects of motherhood, resonating with the newly calibrated sense of levity amidst chaos that the mommy brain possesses (O'Connell 2018). Lighthearted storytelling that reflects maternal experiences provides a much-needed emotional connection, transforming a sales pitch into a friendly conversation.

The maternal brain undergoes enlargement and specialization postpartum, reflecting the remarkable plasticity inherent in the brain's adaptation to ensuring the survival of the next generation. This organic transformation aligns with Darwinian principles and the inherent rhythm of life and death.

The birth experience leaves a lasting impact on the architecture of the mother's brain, imprinting neural pathways driven by maternal aggression and reinforced determination to secure her offspring's safety. These alterations occur rapidly, becoming as distinctive and enduring as a physical tattoo.

The cognitive shift that occurs in mothers is often likened to a highly tuned global positioning system (GPS), with brain areas responsible for focus directing their attention toward their child's safety and well-being. The stability of the nest becomes the focal point around which a mother's world revolves.

Hormones play a critical role in activating the mother's medial preoptic area (MPOA), heightening the brain's reward system. This neurological adaptation parallels the response to intense pleasure stimuli, creating a rewarding loop that brings immense satisfaction to mothers, rivaling the most pleasurable human experiences.

The process of caregiving stimulates brain regions associated with joy and fulfillment, creating a blissful addiction to motherhood. This elevated neural pathway intensity elevates the experience of motherhood to a level that rivals life's most passionate pursuits.

The transition to motherhood represents a renaissance of self, a rebirth into a maternal powerhouse. Mothers navigate their new world with the prowess of a superhero, fueled by oxytocin and displaying resilience and amplified care each day.

According to neuroscience, the brain undergoes significant rewiring in response to motherhood, confirming what many have suspected. Hoekzema et al. (2017) discovered that women experience a reduction in gray matter in certain brain regions, particularly those involved in social cognition and understanding the perspectives of others. However, this reduction does not indicate a loss but rather a refinement that optimizes for the demands of motherhood, likely contributing to heightened empathy and intuitive attunement to the needs of their infants.

These neural adaptations enable a powerful focus on nurturing tasks, placing the well-being of the infant above all else. Rather than perceiving this shift as a diminishment, it should be seen as an extraordinary example of the brain's plasticity, adapting in real time to the crucial task of caring for a vulnerable new life. The implications for marketing are significant. Any brand message aimed at this audience should recognize this focused perspective and appeal to the instinctive drive to nurture and protect.

Contrary to the derogatory term *mommy brain,* implying forgetfulness, scientific research dismantles this myth. Instead of a generalized impairment, studies indicate that mothers may experience enhanced cognitive capacities, such as improved memory for childcare-related tasks and better multitasking. Marketers can leverage this finding by tailoring communication strategies to highlight product features and services that streamline maternal responsibilities and complement these newfound cognitive strengths.

As the maternal brain adapts, so does the maternal instinct. Mothers exhibit an increased sensitivity to environmental hazards and a propensity for vigilant decision-making, particularly regarding products for their children (van den Burg 2020). Marketers should thus shift their focus toward providing comprehensive product information that emphasizes safety and efficacy, meeting the rigorous criteria that now govern a mother's choices.

The intersection of science and storytelling offers a powerful tool for engaging the mommy brain. Narratives that resonate with the reconfigured emotional landscape of motherhood, emphasizing care, interconnectedness, and protection, can forge profound connections. The goal is not simply to sell products to this demographic but to provide affirmations of their maternal instincts and a sense of communal belonging.

Marketers should position their brand as a trusted node within this influential network by fostering experiences based on trust and building communal support, rather than solely focusing on sales. Providing a space for knowledge exchange and empathetic service can naturally integrate the brand into the maternal tribe, becoming an indispensable part of their support system.

One of the defining characteristics of motherhood is the ability to manage emotions effectively, a skill governed by the regulation abilities of the prefrontal cortex (Kim et al. 2018). This emotional equilibrium is a remnant of our ancestors' need to protect their young from hostile environments, requiring negotiation and conflict resolution without escalating to a life-threatening level. In today's world, mothers employ this enhanced regulatory mechanism to navigate diverse arenas, from family tensions to workplace dynamics.

A communications strategy that reflects this inclination toward tranquil interactions is essential. Position the brand as one that empathizes with the pressures of modern motherhood by ensuring that every interaction, from product services to customer support, is streamlined for ease and positivity.

A mother's inclination to avoid unnecessary conflicts extends beyond personal interactions; it also influences her commercial dealings. Brands that provide a serene and hassle-free transactional experience are more likely to earn her loyalty. If there are any shortcomings in a product or service, empower her to raise her concerns without friction. A brand that stands alongside her, navigating the consumer landscape together, will be seen as an ally rather than an adversary.

Recognize the mother's natural aversion to conflict and design customer experiences that are obliging and restorative (Marceau et al. 2015). Problem resolution should be seamless, and the brand's tone should be reassuring. Make the process of lodging a complaint feel more like a therapy session than a battle for her rights.

Throughout human history, altruism has played a vital role in the evolution of communities, and this trait is abundantly evident in a mother's actions. The reward system in her brain is wired to derive satisfaction from acts of altruism, particularly concerning the well-being of her offspring. For marketers, engaging her intrinsic reward circuits through messages of communal value and benevolence is critical.

Embrace this giving nature in marketing approaches. Align the brand's values with causes and contributions that resonate with her altruistic instincts. Campaigns contributing to childcare charities or offering educational tips generate a sense of unified welfare, activating her maternal spirit to act and advocate.

In the digital age, mothers possess resourcefulness that extends into the cyber realm. Modern mothers navigate digital spaces adeptly, leveraging apps and platforms as sources of support and information. Brands that seamlessly integrate into her digital lifestyle by providing utility, connectivity, and empathy can firmly establish a place within her online sphere of reliance (Radesky et al. 2016).

Develop and endorse digital conveniences that respect her time and intelligence. The brand's online presence should offer more than just products; it should provide the warmth of community she seeks. Create apps and platforms that align with her caregiving instincts, magnetizing her digital skills to your brand.

Your marketing approach should be patient and present, delivering the message as a trusted friend holding their hand through each choice. Sell not only a product or service but a partnership that affirms their actions and instinctual path toward the best outcomes for their offspring.

Motherhood is a journey of constant change, and amidst uncertainty, reliability resonates deeply. Be the brand that provides steadfast support in a mother's life, serving as a consistent beacon amidst the ebb and flow of parenting.

Whether through reliable products, dependable advice, or a steadfast community presence, ensure that your brand is synonymous with stability and trust in their ever-changing world.

Communication That Resonates to the Maternal Mind

The phenomenon of maternal bonding is a fascinating reflection of the brain's neuroplasticity, as it adapts its neural connections in response to childbirth and nurturing activities. Research on breastfeeding aligns with broader neurological observations in new mothers, highlighting how the act itself fosters an emotional and physical union that reshapes the mother's mental landscape, merging her sense of self with her child. These intimate care rituals, from breastfeeding to swaddling, enhance the mother's capacity for empathy and reveal the neurological foundations of maternal caregiving.

Motherhood also offers an immersive learning experience in effective communication. Neuroscientific studies suggest that the intimate caregiving activities a mother engages in enable her to decode her baby's nonverbal cues with greater proficiency than she recognizes her own reflections. This bidirectional tutelage in affective fluency is supported by increased gray matter volume in regions like the prefrontal cortex and the amygdala. Through these changes, mothers develop a rich lexicon of emotional responsiveness.

Mirror neurons play a crucial role in the mimicry and empathy observed between mothers and infants. These neurons, which are activated during shared smiles and mutual cooing, strengthen emotional bonds and foster a sense of mutual delight. The insula acts as a neural mediator, translating mimicry into shared positive experiences and facilitating behavioral synchrony in mother–infant interactions.

Depictions of mother and child should emphasize their interconnected existence, showcasing interactions filled with intention and affection. Scientific research emphasizes the importance of genuine connection for the development of an infant's own neural circuitry. Thus, visual narratives in marketing should highlight the mother and child together, immersed in shared joy and purposeful engagement, rather than portraying separation.

Effective communication strategies should reflect the nuances of maternal identity. Acknowledging the protective and attentive embellishments of the mother's new cognitive landscape through storytelling is key. Marketing approaches that authentically embrace and celebrate this transformation resonate deeply and align brand narratives with the enriched emotional palette born of unique neurodevelopment.

Marketing narratives that capture the essence of mother-infant rapport can leverage this neural choreography. Depicting moments like a mother tracing her child's palm or whispering lullabies can evoke empathy and result in positive affect. By reinforcing salient moments of connection embedded in a mother's daily routine, marketing not only reflects but also shapes reality.

To effectively connect with maternal audiences, marketing narratives should embody the unique resonance between mother and child. Visuals and stories that depict the seamless exchange of giggles or the unscripted duet of motherly coos and infantile babbles strike a chord in the hearts of mothers. These narratives are not simply heard but felt, resonating at a frequency that aligns humor and harmony (synchronicity) in unison.

Sensory Enhancement and Swift Decision-Making

The journey of motherhood initiates a remarkable process of neurogenesis, where new brain cells are created and align with the demands of caring for an infant. Samuel Weiss's research at the University of Calgary has shed light on this phenomenon, demonstrating the correlation between the creation of new brain cells and the development of behaviors crucial for motherhood. This expansion in brain architecture allows for stronger connections between a mother and her child.

Motherhood demands exceptional multitasking abilities, an imperative shaped by evolution. Cognitive advancements enable superior navigation and problem-solving prowess as they search for hidden food. This cognitive fortification empowers mothers to divide their attention between various tasks without compromising the care of their children.

The evolutionary blueprint underlying a mother's need for efficiency is profound. The maternal brain is driven by an insistent push to reach its peak resourcefulness. This encompasses understanding the environment, remembering the child's needs and maximizing interaction with the world. A new mother becomes the epitome of efficacy, focusing diligently on critical tasks while diverting attention from nonessential matters to ensure the survival and well-being of both herself and her child.

Neurogenesis in mothers also affects the remodeling of the olfactory lobes, resulting in enhanced olfactory sensitivity that lasts a lifetime. Changes in scents – whether related to the baby, food, or the environment – prompt rapid and often unconscious assessment and responses in the mother's brain (Nehls et al. 2024). This acute sense of smell facilitates bonding with the baby and becomes a crucial part of the mother's interaction with the world, where every odor is diligently decoded for the benefit of her child.

The postpartum period brings about heightened sound sensitivity in the female brain. Within just two days, most new mothers can differentiate their newborn's cry from others, using it as a neural sonar to discern their child's emotional and physical state (Elyada et al. 2015). This auditory augmentation carries significant implications for marketing, highlighting the importance of engaging mothers with soothing and joyful sounds while avoiding auditory displeasure.

Practical application of these insights leads us to envision the marketing landscape as an orchestration of sensory appeal, with the mommy brain serving as the maestro, finely attuned to her child's needs and surroundings. Imagine an aisle tailored to the expertise of maternal senses, with fragrances evoking cleanliness and comfort without overpowering, products emitting gentle sounds rather than noisy calls, and spatial designs that provide a warm embrace.

Merchandisers witness the strategic acumen of the maternal territory, which caters to the mother's need for proximity and practicality. From the arrangement of milk and diapers to the grouping of organic produce and parenting aids, the aim is to facilitate her journey and provide an empathetic nudge toward the registers by creating a sanctuary of essentials. A section impeccably organized, distinct from the labyrinth of endless aisles, appeals to her newfound territorial tendencies. Branding this space and breathing life into it can make it the maternal port in the stormy seas of shopping drudgery.

Hyper Vigilance, Healthy Skepticism, Seeking Trust

The maternal brain exhibits heightened vigilance, particularly concerning safety and well-being. This evolutionary trait is reflected in the shopping and consumption patterns of new mothers, who tend to be meticulous in their caution and conduct extensive research, seeking peer validation before making purchases. Marketing messages targeting this demographic should, therefore, offer transparent and trustworthy product narratives that are evidence-based and endorsed by peers, engaging mothers through a community-based trust network.

It is not just what is being sold but how the story is told. Products and services aimed at mothers should embody narratives that resonate with their rich emotional and cognitive states. This requires a storytelling approach that weaves together themes of security, trust, and care, all of which are deeply valued in the maternal psyche. Mothers are not merely making purchases for consumption; they are investing in elements that will shape their children's lives. Marketers must recognize and respect this crucial point in their campaigns.

Trust and reliability play a significant role in brand relationships, much like interpersonal connections. Neuroscience research suggests that the maternal brain, enriched with oxytocin receptors during this period, becomes more inclined to form strong social alliances and nurture existing bonds. This heightened inclination toward communal affiliations influences a mother's brand preferences, creating deep ties within her trusted tribe of brands.

Hormones orchestrate shifts within the brain that are nothing short of miraculous. Cortisol, often known as the stress hormone, enhances a mother's arousal and vigilance to environmental stressors that may threaten her child. On the other hand, oxytocin underpins the bond between mother and child, fostering attachment and trust. Grasping this intricate hormone symphony empowers marketers to align with the maternal brain's priorities of security, health, and the unbreakable bond with her child.

Motherhood bestows mothers with heightened sensory awareness, an almost superhuman ability to perceive subtle shifts in their baby's environment. It's as if their neurons have developed an intimate connection with the universe, enabling them to sense potential threats or opportunities for their child long before others. Marketers who understand the profound sensory sharpening in mothers can craft product narratives that resonate with a mother's intuitive awareness, aligning their brand messages with the nuances of this sensory superpower.

How Grandmothers Nurture Generations

The "grandmother hypothesis" proposes that humans, unlike other species, benefit from the extended lifespan of women after menopause. Postmenopausal grandmothers play crucial roles in the nourishment, care, and education of their grandchildren, contributing to the survival and thriving of future generations. Neuroscientific insights shed light on the cognitive and emotional faculties that are well-preserved in older women, aligning with their caregiving role.

Neural plasticity, the brain's ability to adapt and reorganize, continues in older women and may be enriched by social interactions and caregiving responsibilities. Grandmothers, as repositories of accumulated cultural knowledge, foster the cognitive development of their grandchildren through storytelling, skill teaching, and problem-solving interactions. These intergenerational exchanges activate brain regions related to social cognition and memory retrieval in both grandmothers and grandchildren.

The emotional connection between grandmothers and grandchildren is influenced by the neurochemistry of attachment, with oxytocin, the "bonding hormone," playing a pivotal role. This hormonal interplay facilitates strong emotional bonds and is correlated with the joy and satisfaction experienced by grandmothers engaged in caregiving.

From a neuroscientific perspective, grandmothers possess cognitive flexibility and problem-solving skills that benefit the educational landscape of their grandchildren (Musil et al. 2019). Additionally, socially active grandmothers demonstrate preserved cerebral functions, particularly in the prefrontal cortex, which is associated with complex social interactions and planning.

Summary

The emergence of motherhood brings about significant changes in the mother's brain, creating a new neural architecture. One notable shift is the reduced activity in the amygdala, traditionally known as the *fear center* of the brain. Studies show that new mothers exhibit decreased amygdala responses to external threats, indicating a shift toward a nurturing-oriented response system. This change aligns with the tend-and-befriend mechanism, which contrasts with the fight-or-flight response.

Additionally, the mother's brain experiences an increase in oxytocin levels, reinforcing her instinct to prioritize her child's survival over her own personal defenses. Prolactin, another hormone associated with parental care, rises to levels necessary for lactation and caregiving. These hormonal changes shape a brain that is less focused on personal safety and more attuned to the vulnerability of the child.

While personal fear diminishes, a new type of vigilance emerges, one that is empathetic and centered around the child. Mothers become calmer in the face of threats, facilitated by their downregulated personal fear responses. This transformation, combined with the intuitive bonds formed through oxytocin, creates an environment conducive to nurturing.

This neurological reorientation offers insights for marketers. Emphasizing tranquility and safety in products and environments can appeal directly to the recalibrated maternal instincts. By showcasing their commitment to child safety, businesses can build trust with mothers who value their child's well-being above all else.

However, it is important to note that the maternal brain is not docile. Alongside the decrease in personal fear, there is an increase in selective aggression aimed at perceived threats to the offspring. This heightened aggression reflects an assertiveness developed by mothers to protect their child. Understanding and leveraging this heightened vigilance and selective aggression can shape marketing strategies that demonstrate a commitment to child safety, resonating deeply with mothers.

GenAi algorithms should have guidelines and guardrails that recognize the following:

- Motherhood brings about profound neurobiological changes in the maternal brain, influencing cognitive, emotional, and social cognition processes.

- The neural adaptations in the maternal brain enhance emotional acuity, empathy, and the ability to decode nonverbal cues, shaping a unique cognitive landscape for mothers.

- Hormonal changes during pregnancy and postpartum play a significant role in reorienting the maternal brain, emphasizing the prioritization of the child's safety and well-being.

- The maternal brain undergoes a renaissance of resilience, optimism, and cognitive flexibility, fostering a profound sense of emotional intelligence to navigate the challenges of motherhood.

- Understanding the neurobiological underpinnings of motherhood is crucial for marketers, enabling the crafting of authentic and empathetic narratives that resonate with the maternal experience.

These key learnings provide valuable insights for marketers seeking to engage with the demographic of mothers, guiding them to create messaging and strategies that authentically align with the neural transformations that occur in the maternal brain. Check for balanced messaging that considers the diversity and complexity of experiences within motherhood.

With this chapter you have gained a solid grasp of the mommy brain in all its variety and uniqueness.

For marketers, you've not only learned the core underlying physical and emotional alterations that have taken place. You're now equipped with specific knowledge and direction about how to reach and win mothers over: everything from the enhanced "GPS" and emotional intelligence at work in her mind to the importance – and the newfound and unique differences – in marketing to her five senses.

Lastly, you now have specific how-tos in crafting the most effective prompts to use with GenAi platforms. As the new and rapidly evolving language of our times, your mastery of this technology will be ever more critical to your success in an AI-driven world.

CHAPTER 6

The Middle-Aged Brain

T urns out, it's true. The midlife crisis is indeed a thing.

They say that every crisis is an opportunity. Most people in their forties would acknowledge that it is an opportunity for a lot of mischief to happen in life.

And neuroscience knows why.

In this chapter you will learn much about the "how and why" of the middle-age mind, when life changes can be profound – and occasionally perplexing.

Drops in critical neurochemicals in the brain can trigger everything from a loss of sleep to a drive to a skydive. But they also harbinger positive needs – seeking human society and the amazing power of touch to hoist our spirits. Turns out that a hug is more than nice – it's a driver of greater well-being.

You're about to be confronted with *crystallized intelligence*. Read on to find out why.

Dopamine Drops and Midlife Mischief

Neurotransmitters play a vital role in the communication between neurons in our brains, orchestrating the intricate symphony of thoughts, emotions, and actions that define our human experience. One neurotransmitter, dopamine, often referred to as the "feel-good" chemical, is closely linked to our pleasure and reward circuitry. Engaging in new experiences, enjoying our favorite foods, and achieving success all contribute to the production of dopamine, resulting in a sense of elation and satisfaction (Lembke 2021).

However, research suggests that as we enter middle age, there is a decline in neurotransmitter levels, particularly dopamine, which could explain why activities that once brought us joy may start to feel less fulfilling. It is as if someone turned down the volume on the soundtrack of life. While these findings are nuanced and involve various factors, they significantly contribute to our understanding of the neurological changes that occur as we age.

This decrease in dopamine production during midlife is not solely about fading joy; it is about our brains seeking balance. It nudges us toward behaviors that might replenish our diminishing dopamine levels. This pursuit of equilibrium is often referred to as the "midlife crisis," where individuals seek dopamine boosts through novel experiences. Late-night internet searches for sports cars and adventurous activities represent the brain's attempt to reignite that waning dopamine surge.

However, not all midlife transitions involve daredevil territory. Many individuals seek rejuvenation through more subtle means, such as developing new hobbies, exploring travel, or fostering new relationships. These pursuits demonstrate the brain's adaptability and show that sparks of excitement can come from unexpected endeavors.

Understanding the quest for dopamine in the middle-aged brain can inform marketers seeking to connect with this demographic. Recognizing the "why" behind the thrill-seeking and novelty-chasing nature of this age group allows brands to craft turnkey messages. Products and services can be positioned as not just purchases but as gateways to adventure, symbols of rejuvenation, and catalysts for reinvention.

For example, a travel agency specializing in curated adventures for middle-aged thrill-seekers could create an ad campaign that goes beyond detailing the destination. Instead, it could weave a narrative of rediscovery and the exhilaration of stepping outside one's comfort zone, while providing a sense of familiarity and safety that individuals at this age may desire.

In the end, marketing that reflects the evolving script of life and respects the underlying neurological currents of its audience can establish a powerful connection. It involves presenting a product or experience as a companion in the journey of reinvention and a confidante in the pursuit of joy. Because even as we age, our brains never stop longing for the dopamine dance.

Desperately Seeking Serotonin – World of Color Transforms to 50 Shades of Gray

Serotonin, the neurotransmitter synonymous with feelings of well-being and satisfaction, serves as emotional comfort food. Just like cradling a mug of hot chocolate on a chilly evening, serotonin works its calming magic. It plays a vital role in regulating mood, appetite, and sleep, among other functions. When serotonin levels decline, it's akin to losing that nurturing cup of cocoa, leaving us devoid of life's cozy reassurances.

The effects of a serotonin dip extend beyond a mere absence of comfort; they manifest as tangible challenges at both physiological and psychological levels. Disrupted sleep, a hallmark of midlife changes, may occur. There may also be a loss of drive, transforming tasks that were once approached with enthusiasm into mundane endeavors. Life loses its luster, casting a shadow of existential grayness over everything.

In response to this internal shift, individuals may exhibit introspective and isolative behaviors. As serotonin levels decline, the desire for social engagement diminishes, leading many to turn inward (Żmudzka et al. 2018). Relationships that flourished on shared satisfaction may endure cold periods as individuals grapple with their internal sense of disenchantment, struggling to articulate what they feel is missing.

This neurochemical turmoil, marked by a drop in serotonin and dopamine, brings about the proverbial midlife crisis. It is characterized by sudden shifts in life decisions, such as career changes, relocations, and even relationship dissolution. The impact is not limited to the individual alone but ripples through their entire ecosystem, shaking the foundations of established families and communities.

Addressing this biochemical crossroads requires both awareness and action. How can marketers leverage this understanding to benefit those navigating this tumultuous terrain? Here lies an opportunity to approach this demographic with products and strategies that offer solace and excitement, tailored to their neurochemical needs.

For marketers, the key lies in recognizing that a sense of adventure can be a powerful remedy for the serotonin-deficient soul. Encouraging activities or products that inspire fulfillment or reconnect individuals with themselves can deeply resonate with those longing for their lost "hot chocolate" moment.

There is potential for a campaign that not only promotes products but also suggests wellness experiences that naturally boost serotonin levels. Spa visits, meditation retreats, or gardening kits can be positioned as avenues to reclaim inner harmony. The goal is to paint a picture where the product transcends mere commodity status to become a stepping stone toward rediscovered inner harmony.

Consider, for example, a line of teas enriched with natural substances known to support serotonin levels. An advertisement for such a product could feature comforting, homely scenes that exude warmth and familiarity, speaking directly to the deep-rooted yearning for coziness and comfort experienced by those going through a serotonin slump.

In the end, midlife should be viewed as a transition that should be handled sensitively and understandingly rather than just as a crisis to be lamented. When marketers are aware of their target audience's neurochemical story, they may provide more than just products – instead, they can act as a key to opening the next happy chapter in life.

Daily Novelty – Inject Excitement and Mojo into Midlife

The relationship between dopamine and novelty is intricately linked to neural circuits like the mesolimbic pathway, which has evolved to reward novel-seeking behaviors. This circuitry ensures that seeking new experiences remains a vital part of our survival instincts by encouraging exploration and learning. In our modern lives, this translates into the need for daily variety to prevent the monotony of routine.

Even small deviations from our everyday routine can trigger the release of dopamine and bring a sense of anticipation and psychological well-being. From trying a different route on our commute to experimenting with a recipe or engaging in a fresh conversation topic, these "fortune cookie moments" sprinkle unexpected sweetness into our days and act as cognitive refreshers.

The human affinity for novelty is not merely a fleeting desire but a deep-seated urge rooted in our neurobiology. When we encounter something new and unconventional, our brain's reward circuitry is activated, leading to the release of dopamine, a neurotransmitter associated with pleasure and motivation (Voruganti et al. 2007). This release of dopamine creates a sense of enjoyment that often accompanies novel and rewarding experiences, indicating that our preference for novelty is not solely influenced by culture but is an inherent part of our neural makeup.

Considering the adaptability and plasticity of the human brain, it is unsurprising that our neurochemical responses are finely attuned to the element of surprise. From an evolutionary standpoint, this biological inclination toward novelty likely motivated our ancestors to explore and discover, thereby increasing their chances of survival. In contemporary times, this translates into a constant quest for engaging and novel experiences, echoing our exploratory heritage at a physiological level.

Understanding the science behind novelty elucidates the allure of dopamine. Experiences that are novel or varied can activate the mesolimbic pathway, a neural pathway that specifically rewards the discovery of new stimuli. This evolutionary wiring allows even mundane routines to become opportunities for the brain to seek dopamine "hits" in our everyday lives.

Given our inherent predisposition for novelty, the daily occurrences in life present numerous opportunities to introduce small yet meaningful changes that keep our dopamine levels elevated. Engaging in a new hobby, trying different cuisines, or modifying our regular commuting routes can act as catalysts for cognitive stimulation and emotional uplift. These seemingly insignificant alterations can serve as a neurological reset, combatting the monotony of familiarity.

In a café setting, where individuals gather for their mutual love of caffeine and conversation, each table could host conversation starter cards that provide potential dopamine delights with each flip. This clever approach not only satisfies the brain's thirst for novelty but also capitalizes on our inherently social nature, which influences our daily behaviors and sense of satisfaction.

To put these insights into practice, we can design personalized messaging that offer daily doses of novelty. This could include desktop gadgets with inspirational quotes, meal kits with surprise ingredients, or digital platforms suggesting randomly generated daily tasks. Each day presents an opportunity to satisfy our brain's craving for curiosity, infusing playfulness and joy into the ordinary.

These insights offer marketers a canvas full of potential. Products and services can be positioned to enhance everyday life by providing both novelty and nurturing. A phone app encouraging random acts of kindness among peers or a service delivering handcrafted notes from strangers could effectively fulfill this niche. By tapping into the brain's desires through humor, warmth, or bewilderment, marketers can forge deeper connections with their audience.

Indeed, GenAi itself becomes a powerful tool to satisfy the urges for novelty and comfort. The "playgrounds" of GenAi engines offer powerful ways to be pleasantly surprised by what the engines can create, and more importantly co-create. The allure of GenAi engines to those in the tender ages of 40 to 50 is real and tangible.

Touch, Hug, and Gratitude – Serotonin Boosts without Pills

The language of human touch speaks directly to our neurology, fostering comfort, connection, and well-being. When we embrace another person, serotonin, a powerful neurotransmitter, floods our brain, creating a pacifying effect and enveloping our neural landscape (Hensler 2010). This neurochemical communication not only strengthens bonds but also brings tranquility akin to the soothing sensation of sipping hot chocolate on a chilly evening, enveloping our psyche in warmth.

The power of touch extends beyond humans, affecting our animal companions as well. Petting a pet triggers a serotonin cascade, inducing calmness and reducing stress. This phenomenon is supported not only by subjective feelings but also by empirical studies measuring blood pressure and heart rate in both humans and animals (Watts et al. 2012). It seems that the act of touch distills the essence of comfort into a tactile moment, providing a biochemical reassurance that all is well.

This "serotonin fix" through physical connections highlights a fundamental principle: Our minds crave joy in small packages, everyday moments that ignite sparks of happiness. Engaging with life's simple pleasures, whether through a heart-warming hug or the excitement of a new experience, can replenish our inner reserves of serenity and satisfaction. Regularly integrating these comforting and refreshing moments into our lives can help alleviate symptoms of anxiety and depression, enhancing overall life appreciation and satisfaction.

In our fast-paced lives, we often overlook these moments of joy. Yet, by intentionally seeking them out, we can multiply their neurological benefits. Creating opportunities for serendipitous pleasures is like setting up rendezvous spots for chance encounters with happiness. It involves rediscovering the allure of our daily lives, which are teeming with hidden treasures that boost serotonin levels.

By merging the realms of science and the art of living, we can cultivate a life where serenity is not just a pursuit but a natural state, nurtured alongside an enthusiastic embrace of life's countless little joys. Brands and products that align themselves with these delights, offering not just goods or services but experiences, can resonate with a public constantly seeking their next serotonin surge.

Shifting our focus to gratitude, it is widely recognized that acknowledging our blessings triggers the release of serotonin, reinforcing feelings of well-being and contentment (Sirgy & Sirgy 2020). This presents an opportunity to integrate neurological insights with interactive consumer engagement. For example, a mobile application that prompts users to record moments of gratitude can offer a double benefit: a serotonin boost through the act itself and a dopamine surge through the gamified digital interface.

GenAi imagery and language that emphasizes close contact, and hugs provides more than linguistic satisfaction, it provides the needed therapy for consumers who are serotonin deprived.

Light at the End of the Tunnel – Brain Shrinkage to the Rescue

Brain shrinkage, once regarded as an unwelcome sign of aging, has recently been subject to reevaluation. The decrease in brain volume, particularly in the prefrontal cortex and hippocampus, was previously associated solely with cognitive decline. However, emerging research from neuroscience suggests that this process and its impact on the limbic system, which may also involve a reduction in emotional volatility, has the potential to lead to increased emotional stability and improved interpersonal relationships. This brain shrinkage occurs toward the end of the "midlife crisis," at around 50 years of age. So there is hope after 10 years in darkness.

As individuals age, significant changes occur in the brain's limbic system, which regulates emotions. The amygdala, responsible for fear and aggression, shows decreased activity in response to negative stimuli in older adults compared to younger individuals, potentially explaining the mellowing of emotions that often accompanies aging (Jacques et al. 2010). This diminished emotional volatility may contribute to the observed increase in positive social interactions among older adults.

Not all aspects of brain function deteriorate during this period. Certain cognitive abilities, such as vocabulary and knowledge, continue to thrive or remain stable. Within this context, the term *brain shrinkage* may not solely indicate loss but rather represents a pruning process that prioritizes efficiency over capacity. Similar to sculpting a refined masterpiece from a block of marble, the brain undergoes a transformation aimed at optimizing neural connections.

This shift toward neural efficiency aligns with the concept of *crystallized intelligence* – the ability to utilize knowledge acquired throughout a lifetime. While *fluid intelligence* or the capacity to process new information may decline, crystallized intelligence is maintained or even enhanced, serving as a compensatory mechanism for age-related changes in brain structure.

These findings provide opportunities for proactive adaptation rather than resigned acceptance. Neuroscientists commonly emphasize the benefits of habitual mental and physical exercise in bolstering brain health, potentially counteracting volume loss by promoting neuroplasticity, the brain's ability to form new connections throughout life (Phillips 2017).

Applying these scientific findings, one can consider the cultivation of new skills and hobbies as a form of mental cross-training, bringing vitality to synapses. Learning a second language or a musical instrument in midlife, for example, may serve as a cognitive fountain of youth, enhancing brain health and broadening sociocultural horizons.

Furthermore, as our neural landscapes evolve, our social connections can evolve too. The stability and warmth that accompany the neural changes in midlife reinforce the importance of deepening existing relationships and fostering new connections (Hagestad 2018). Engaging in community activities, volunteer work, and mentoring opportunities can harness this phase of life, characterized by enhanced relational skills, and contribute to personal fulfillment and social cohesion.

It is crucial for marketers and talented storytellers to take note: This generation is ready for a renaissance, not a decline. Products and advertising campaigns ought to acknowledge this transition, showing the way from the chaos of midlife to a tranquil twilight full of opportunities. This story arc, which moves from storms to sunshine while capturing and engrossing the viewer, is driven by the confidence of science and the comfort of serotonin. The storm clouds and maladies that appear at 40 seem to magically mellow and disappear at 50.

Hope dawns again, and the second golden age of the brain begins.

Now you've cracked the code of why our basic behaviors and our outlook on life undergo such significant changes at midlife.

You have gained insight into how best to market to that middle-aged mind. You've also learned that, despite certain physical and emotional declines, there is much to anticipate and much to celebrate at life's midpoint.

Moving forward from our more tumultuous early years, the mature brain achieves a more mellow state. Wisdom gained.

And you have a clear path to market to the middle-aged mind, with concise guides to creating the most effective prompts for GenAi platforms.

CHAPTER 7

The Senior Brain and What It Prefers

For a full generation, our society has been fascinated by the mature and senior generation – its music, its economic power.

It has been not only the largest cohort, but arguably the most impactful.

Modern neuroscience has plumbed the depths of the senior brain (Knight 2018) and, as this generation matures, provided insights that can guide marketing campaigns to success.

In this chapter you will gain a broad understanding of the dynamics that drive the senior mind, and their implications for everything from new product introductions to brand loyalty, and more.

You will learn the importance of memories, music, fragrance, clarity, simplicity, and a whole host of other factors that should be called upon to market to the senior generation. You will also learn how to harness the power of GenAi to connect meaningfully with seniors.

While most of the effort is currently spent today on figuring out how to talk to "Gen-Alphabet Soup," not enough effort is spent on understanding how to understand, appreciate, and communicate effectively with those that gave birth to the young ones – seniors.

Incidentally, seniors have incredible spending power. GenAi allows us to address this "age gap" and find respectful and potent ways of communicating with this incredible audience. This chapter in a way is dedicated to all our parents and grandparents, without whom we would not be where we are today.

Keep It Simple – and Focused

With an ever-expanding marketplace and the advent of new media platforms, understanding the psychological mechanisms driving consumer behavior remains at the forefront of effective marketing. Successful strategies often rely on a marketer's ability to navigate the complex landscape of cognitive processes, particularly as it pertains to attention and memory. These cognitive dimensions play pivotal roles in how potential customers engage with advertisements and retain product information.

Attention can be likened to the beam of a lighthouse, constantly scanning the environment for noteworthy information amid a sea of sensory stimuli. It's selective by necessity, as the sheer volume of information we encounter daily would overwhelm our cognitive capabilities were we to attempt to process it all. As we age, this selection process doesn't necessarily deteriorate in focus – the lighthouse's beam remains sharp – but its ability to ignore the irrelevant – the distracting waves and seagulls – diminishes (Commodari et al. 2008).

Peripheral distractions – those that fall outside the focus of our attentional beam – can more easily penetrate the cognitive barriers of older adults. This increased susceptibility to distraction can implicate greater challenges in message retention, which is crucial in ensuring that marketing efforts leave a lasting impression. It is not the presence of the distraction itself that is the issue, but rather the decreased efficiency in suppression mechanisms – akin to a lighthouse with a dimmed surrounding darkness, allowing more than just the intended focal point to shine through.

Neuroimaging studies illustrate that the activation patterns in the brains of older adults differ from those in younger individuals when engaged in tasks that require attentional focus. The networks that support suppression of irrelevant information show reduced activity, which can lead to higher levels of interference by extraneous stimuli. This contrasts with the sustained activation of the networks responsible for maintaining focus, suggesting that while seniors can concentrate on the intended message, they are often fighting an uphill battle against a backdrop of distraction.

The implications of this for marketing are manifold. For one, the utilization of clear and concise messaging becomes paramount. Cluttered visual layouts, complex sentence structures, and the presence of nonessential information act as fodder for distraction. To optimize the attention of older consumers, marketers should present information in short, digestible bites that do not require heavy cognitive lifting to understand.

While simplicity offers clarity, it must be balanced with emotional resonance. Affective content has been shown to enhance memory recall, particularly in older adults. This is due to emotion's ability to focus the mind and act as a catalyst for deeper cognitive processing. Therefore, an ideal marketing message combines simplicity with an emotional hook that not only attracts attention but also heightens the potential for message retention.

The inclusion of narratives within marketing materials has shown promise as a method of leveraging emotion to both capture and maintain attention. Stories, particularly those to which mature audiences can relate, have the potential to engage an audience on an emotional level while simplifying complex information into digestible and memorable vignettes (Tessitore et al. 2005). For marketers, this means weaving product details into a relatable human context rather than listing features and benefits in isolation.

However, as marketers, we must also be wary of information density. Pacing is key; too much too quickly can overwhelm the cognitive systems responsible for both attention and memory. Integrating pauses into marketing messages, allowing for moments of reflection and consolidation, can aid information processing and enhance the likelihood that the information will move from short-term to long-term storage.

Finally, there's the format to consider. A multimodal approach that employs both visual and auditory elements capitalizes on the brain's ability to integrate information from various senses, supporting better recall. For an older audience, enhancing auditory messages reinforced with visual cues can reduce the cognitive burden associated with processing them in isolation and can mitigate susceptibility to distraction (Heinrich et al. 2016).

Marketers who design campaigns that are aware of and cater to the attentional and memory capacities of older adults will be better positioned to engage this audience effectively. By focusing on simplicity and clarity, leveraging emotional resonance, narrating relatable stories, aligning with brand familiarity, pacing content appropriately, and utilizing multimodal delivery, we can enhance the appeal of our marketing messages and, ultimately, the likelihood that they will resonate with and be remembered by older consumers.

GenAi with its "temperature settings," can wander, hallucinate, imagine, and generally create interesting content, but that just does not cater to the senior brain. Being direct and simple, with short, structured content free of distractions, should be guiding principles for GenAi prompts and algorithms. Understanding the senior mind requires image GenAi to be sensitive to font sizes, colors, contrast, and the general frailties of this incredible generation. The power of GenAi allows immediate and even real-time personalization of messaging and imagery knowing the age of the audience. This enables senior brains to feel youthful, powerful, and competent – enjoying that aging has not robbed them of any of the faculties and powers they once possessed. GenAi can serve as the great age equalizer.

Positive Framing Only

Studies demonstrate that older adults are more prone to favor positive information over negative in their attention and memory processes (Wirth et al. 2017). This phenomenon, known as the *positivity effect*, indicates that the aging brain naturally gravitates toward positivity, potentially as an adaptive measure to maintain emotional well-being.

This inherent preference for positive stimuli can alter an individual's attention network, affecting what information is noticed and focused on. In younger adults, negative information often captures attention more readily due to its potential threat value. However, as one ages, the amygdala – a brain region associated with emotional processing – exhibits a stronger response to positive stimuli and a dampened response to negative counterparts. This suggests a neurobiological shift toward positive selective attention in older adults.

Like attention, memory consolidation is similarly influenced. Older adults tend to remember emotionally positive content better than negative or neutral content. This selective memory bias reinforces the psychology of positivity in marketing toward the mature market. For brands, this requires crafting messages that highlight the positive aspects of products and services rather than focusing on fear or avoidance of negative outcomes.

This notion is supported by research into motivation and behavior change. According to the socioemotional selectivity theory, as aging people perceive a shrinking time horizon, they prioritize emotionally meaningful goals (Carstensen 2021). Emphasizing the enriching experiences or emotional rewards that come with using a product can align with the underlying motivations of mature consumers and have a greater impact on their perceptions and decisions.

Translating this into actionable strategies, marketers should eschew the stereotypical "elderly" themes of decline and instead focus on aspirational and uplifting content. Seniors are not a monolithic group and often bristle at being reminded of their age. Instead, they respond better to nodal age, where the marketing focuses on life stages and the shared values or experiences common to a particular cohort. By understanding these values, marketers can create content that resonates on a positive, emotional level.

The design aesthetics of marketing materials should also reflect a buoyant and optimistic tone, employing bright colors, clear imagery, and fonts that are both legible and convey a sense of warmth. This not only assists in readability for those whose vision may not be what it used to be but also unconsciously embeds a feel-good reaction to the materials.

Simplicity is key; the material should be uncluttered and straightforward. Given the preference for positive information, it is essential that the core message – the "sunshine," so to speak – shines through without "clouds" of unnecessary complexity to obscure it. Clear, coherent, and catchy phrases can be more easily recalled and can linger in memory, bringing a smile when the product or brand comes to mind.

In application, this knowledge dictates that to effectively market to seniors, brands must position themselves not merely as a product or service provider but as a harbinger of positive experiences, well-being, and joy (Gau 2019). This could take the form of advertisements that celebrate the accomplishments and active lifestyles of seniors or communications that focus on the enjoyment of life with friends and family, brought together by the marketed product.

Ultimately, the message for GenAi algorithms is clear: to resonate with the mature mind, wrap your brand in a package of positivity. A positive message is not only more likely to be noticed and remembered; it resonates with the emotional desires of older consumers for content substance and emotional depth. By acknowledging and leveraging the power of positivity, marketers can craft campaigns that both capture attention and also earn a fond place in the memory of their audience.

GenAi can help in simplifying complex information into easier-to-understand messages without losing the essential details. This is particularly beneficial for seniors who might find overly technical or jargon-heavy language challenging to comprehend.

Scented Memories

Few sensory experiences are as evocative as the power of fragrance. Neuroscience reveals a close link between our sense of smell and the neural pathways associated with memory – a relationship that deepens with age. The olfactory bulb, responsible for processing odors, has direct connections to the amygdala and hippocampus – key areas involved in emotion and memory. Remarkably, this sensory route is particularly short, granting the scent a potent capability to trigger vivid recollections.

The phenomenon, known as the *Proust effect*, aptly underscores the intimate association of smell and memory. The term arises from Marcel Proust's literary description of how a whiff of madeleine cake evoked a profound sense of nostalgia. Scientifically, this could be attributed to the fact that the hippocampus, the cradle of long-term memory consolidation, lies a mere two synapses away from the olfactory nerve receptors. This proximity means scents can effectively and quickly access memory stores, arguably more so than other senses (Green et al. 2023).

For the senior mind, these connections appear even more robust. As we age, certain cognitive functions, such as those involving the frontal cortex, may diminish, but the primal olfactory pathways often remain unscathed. Studies show that odors can act as cues to awaken memories, and in doing so, can even enhance the clarity and detail of recalled events. Seniors, therefore, may experience a sharper, more pronounced revival of memories when stimulated by familiar scents.

Aging also brings changes in emotional processing, with older adults showing a preference for positive over negative memories. This positivity bias may interact with olfaction, meaning that marketers can strategically employ scents known to evoke a positive emotional response in people. Scents like vanilla and lavender, commonly associated with comfort and calm, could be particularly effective in marketing contexts targeting the mature demographic.

Translating this science into GenAi guidance means integrating fragrance in ways that are subtle yet impactful. Consider the environment in which products are presented. Spaces infused with a pleasant fragrance can lead to longer browsing times and more positive customer evaluations, provided that the scent is congruent with the product offering. Retailers need to be mindful of scent intensity and type, given that overpowering fragrances can have the inverse effect, leading to a sensory overload or misattribution of emotion.

Furthermore, the careful selection of scents used in GenAi outputs could also yield higher levels of customer engagement and brand recall. For instance, a scented catalog or a fragranced direct mail advertisement could trigger reminiscence and strengthen brand-customer associations. The key is to ensure that the fragrance aligns with the brand ethos and messaging, creating an aromatic narrative that complements the visual and textual elements.

Designing marketing materials with older audiences in mind also necessitates an understanding of age-related changes in the olfactory system. While core olfactory pathways remain relatively resilient with age, the sensitivity to certain scents may diminish slightly (Doty & Kamath 2014). Accommodating these changes without overwhelming the olfactory system is a delicate balance – one that requires precise calibration of the intensity and familiarity of the olfactory cues used.

GenAi can analyze individual preferences, histories, and emotional responses to different scents to develop personalized fragrances that resonate with or evoke positive memories for seniors. This customization could enhance emotional well-being and recall of cherished memories. In addition, even in the simple craft-ing of messaging for boomer brains, invoking *flowers* and *spices* as trigger words creates neurological resonance and increases effectiveness.

Songs and Jingles of Youth – Musical Nostalgia

Music, with its rhythm, melody, and harmony, can serve as a profound mnemonic device invoking powerful emotions and vivid memories. In the exploration of the senior mind, it emerges as a melody playing upon the strings of nostalgia and remembrance.

Neuroscientific studies suggest that our musical preferences and peak emo-tional experiences with music typically form during adolescence and young adulthood, particularly between the ages of 14 and 22 – a period termed the *rem-iniscence bump*.

The limbic system, a neural network involved in emotional responses and memory processing, is uniquely activated by music. Within this system, the hippocampus is critical for long-term memory formation, while the amygdala attaches emotional significance to memories. During our formative years, the limbic system undergoes significant development alongside hormonal changes that heighten emotional intensity.

Indeed, these neural dance partners – the hippocampus and amygdala – couple more tightly in response to music from our youth. As these memories are laid down during an emotionally charged period, they tend to be preserved with remarkable fidelity. Thus, for the aging population, familiar tunes from their youth can "re-tune" the brain to its youthful timber and evoke rich, multisensory autobiographical memories (Belfi et al. 2016).

Musical nostalgia, as some experts call it, isn't simply about remembering a song. It's about reliving the moments and emotions that the song encapsulates. For the mature mind, hearing a melody from the past can often trigger a vir-tually unparalleled sense of presence and clarity about a moment long passed. Researchers have also found consistent activation in the cerebellum, which is implicated in emotional processing and voluntary movements like dancing, sug-gesting music's propensity to make us want to move to the beat, a universal reac-tion regardless of age.

Brands that harmonize their messaging with the soundtrack of the senior mind can see amplified engagement. Incorporating music from the "reminiscence bump" era into commercials, for example, can create an earworm that links a product with a fond memory, allowing for a deeper emotional bond to form bet-ween brand and consumer.

Product presentations accompanied by music familiar to seniors can be much more compelling. For instance, a car commercial underscored by music evoking the era of the vehicle's heyday can resonate with senior customers recalling the freedom and adventure of their youth.

It is essential for marketers to strike the right note and volume. While youth-centric tunes can rev up nostalgia, blasting them as though the senior crowd is still in their raucous concert-going days might overplay one's hand. A touch of finesse – a hint of those iconic riffs, a subdued baseline – captures the spirit without overwhelming the senses or appearing patronizing.

Altogether, the song remains the same: Music is an anchor to our past, a key that unlocks doors in our memory we thought were sealed. For marketers targeting the mature mind, understanding this connection holds the potential to craft campaigns that resonate on a deeply personal level. The rhapsody of remembrance is a powerful tool, and when played artfully, it harmonizes brand stories with the audience's life stories in a degree of delightful recall.

GenAi can generate playlists, and sonic branding can be tailored to evoke positive memories, improve mood, or provide comfort, catering specifically to the tastes and therapeutic needs of seniors.

Leveraging the connection between music and memory, GenAi can help identify tracks or musical styles that are most effective at triggering reminiscence in senior brains. By analyzing emotional responses to different types of music, GenAi can recommend music that will help manage stress, anxiety, or depression. This can be particularly useful in creating calming environments for seniors, especially those in assisted living facilities or undergoing medical treatments.

Grandparents and Grandchildren – A Special Bond

The allure of youth and vitality is a powerful force in the human psyche. Neuroscience reveals that the senior brain is particularly attuned to the energy and zest of younger generations. This magnetic attraction may be underpinned by an evolutionary mandate to ensure the success and well-being of subsequent generations. Dopamine, a neurotransmitter associated with pleasure and reward, plays a crucial role in this dynamic. The senior brain may indeed experience surges in dopamine when interacting with the young, lighting up reward circuits associated with positive reinforcement and emotional satisfaction.

One of the most emblematic relationships showcasing this phenomenon is the bond between grandparents and their grandchildren. The intimacy and joy drawn from these interactions are more than anecdotal; they are grounded in the enrichment that multigenerational engagement provides. For the elderly, it's not just a sentimental tie but also a biological imperative. Studies point out that older adults who frequently spend time with younger family members tend to have sharper cognitive functions, suggesting that these relationships can help to keep their brains engaged and active.

This cognitive coalescence is mirrored in the so-called *grandparent hypothesis,* which proposes that the presence of grandparents in a family unit has been beneficial for the survival of the group across human evolution (Lieberman et al. 2021). This hypothesis is supported by research showing that grandparents' assistance in caregiving not only benefits the ontogenesis of the grandchildren but also offers positive feedback loops to the seniors' health and longevity.

The increased time spent with grandchildren or young individuals can have profound impacts on the mental and emotional health of seniors. Social interactions with the young could lead to reduced risk of depression, a common challenge faced by the elderly population. Studies have indicated that meaningful relationships across generations not only improve seniors' emotional well-being but can also foster a sense of purpose and contribute to their overall sense of belonging within the family and society.

Surprisingly, the benefits don't solely gravitate around intensive cognitive exercises. Even leisure activities shared between seniors and the younger generation can instigate neurological stimulation. Simple acts like playing a game, storytelling, or even walking together can induce mental enrichment in older adults, providing cognitive and sensory stimulation that is both necessary and therapeutic for the maintenance of brain health.

As some may whimsically say, "You're only as old as you feel," and the science seems to agree. Engagement with the young can promote a subjective sense of youthfulness in older individuals. It's not so much about denying the aging process as embracing a state of mind that fosters youthful attitudes and perspectives, which is beneficial to both mental acuity and physical health in the elderly populace.

Integrating these neuroscientific insights, businesses and marketers can form strategies to craft messages that emphasize vivacity and intergenerational connectedness. Brands can build campaigns that celebrate the role of seniors within families and communities, invoking images and narratives that feature them sharing joyous moments with younger individuals. These portrayals can trigger similar neurological responses associated with pleasant intergenerational interactions.

Marketing initiatives could further offer products and services that facilitate these connections, such as family-friendly travel packages or experience vouchers. Organizations may also consider creating community platforms that encourage volunteering and mentorship opportunities that would embrace and harness the power of intergenerational bonds.

On a societal level, engagement across generations can be fostered through programs and events that bring together seniors and youth. Schools might collaborate with elder centers to set up "adopt-a-grandparent" schemes, or urban planners could design parks and communal areas that are equally inviting to all ages, encouraging spontaneous intergenerational interaction. Such initiatives could create everyday bridges between age groups, contributing to a more cohesive and healthy society.

In conclusion, the symbiotic bond between seniors and the younger generation is one that carries mutual reward, deep-rooted in our neurology and psychology. It behooves us to cherish and harness this dynamic, celebrating the timeless crackle of vitality that ignites when age is just a number and joy springs eternal. Harnessing this aspect of humanity can lead to healthier, happier lives across the lifespan.

Marketers can employ GenAi to craft stories or narratives that feature intergenerational themes, showcasing the value of connections between seniors and their grandchildren. Marketers can also utilize GenAi to analyze data from both seniors and younger demographics to identify overlapping interests or activities that could serve as a bridge for connection. Marketing materials can then focus on these shared interests, promoting products or services that facilitate shared experiences, such as games, educational apps, or hobby-related items.

Pragmatism, Realism, and Acceptance

In the grand course of time, the senior brain adapts its tempo, favoring reality over romanticism. Through the gradual process of aging, the brain's architecture undergoes significant alterations, not least within the realms of the prefrontal cortex and the limbic system. The prefrontal cortex, an instrumental executive functioning area, often takes on a more dominant role as we age, instilling a sense of realism in decision-making processes (Denburg & Hedgcock 2015). This shift toward a pragmatic outlook may reflect an evolutionary advantage, fostering adaptive strategies that prioritize survival and well-being in the latter stages of life.

Concurrently, neuroimaging studies suggest a downscaling in the limbic system, a constellation of brain structures pivotal in regulating emotions. The shrinkage of regions like the amygdala is accompanied by a diminishing reactivity to emotionally charged stimuli. Consequently, older adults tend to exhibit a tempered response to life's emotional rollercoaster, settling into a serene realism that prioritizes practical outcomes over idealized expectations.

Such neurobiological evolution begets a nuanced worldview, stripped of rose-colored lenses. The senior brain regards life events – whether they be of love, loss, or the myriad others – with a clear-eyed perspective (Levitin 2020). The romantic notions that once may have shaped one's views in younger years are now cast aside in favor of a more measured, sagacious approach.

It is important to note, however, that this transition does not equate to a loss of emotion or humanity. Rather, it illustrates an admirable capacity for adaptation and the adoption of coping mechanisms that safeguard emotional health amid the natural declines of cognitive function that accompany aging. The senior brain, thus, is not deficient but different, reconfigured for resilience.

Seniors typically approach significant life decisions with a pragmatic air, weighing the pros and cons with analytical detachment that can lead to more satisfaction and less regret.

Moreover, the seasoned wisdom of the aging brain impacts social dynamics uniquely. With the fading desire for emotional highs and lows, seniors often find themselves in the role of stabilizers within their families or communities. They're less likely to be drawn into drama, instead offering a steadying presence that can be both comforting and enlightening to those around them.

Armed with the confidence of their life experiences and the freedom afforded by their realistic outlook, many seniors are known to engage in lighthearted banter and social merriment with a sort of cheeky wisdom.

Marketing and communication strategies, therefore, would do well to tap into this blend of pragmatism and mirth. Campaigns that acknowledge the grounded wisdom of the mature mind, perhaps with a twinkle of gentle humor about the realities of aging, will find a receptive audience. For instance, a portrayal of seniors making savvy consumer choices, or navigating technology with a blend of humor and deftness, can resonate well with what we understand about the mature brain's tilt toward realism and functionality.

It's certainly a balancing act – creating content that speaks to the realistic side of the mature mind without veering into the realm of stereotype or condescension. A well-crafted message acknowledges the blend of humor and practicality that characterizes the wisdom of the aging population. Marketers can convey respect and appeal by featuring stories that celebrate the clear-eyed insights gained through a lifetime of experience.

The senior brain's shift toward a more realistic and practical view of the world is a fascinating evolution that underscores the adaptability of the human mind. By understanding and honoring this cognitive transformation, we can celebrate and engage with our senior population more effectively, integrating their insights, humor, and sagacity into a shared narrative that enriches us all.

GenAi can be guided to create that which reflects the wisdom and life experiences of the senior audience. This could involve sharing stories or insights that echo the lessons learned through a lifetime, making the content more relatable and engaging for seniors.

GenAi can help identify and highlight product features that are endorsed by experts or that incorporate expert knowledge. This approach taps into the senior audience's respect for authority and experience, positioning the brand as a wise choice.

We Mattered: Nostalgia, Accomplishment, Recognition, and Acknowledgment

The tapestry of our lives is rich with accomplishments and memories that define who we are. For the senior members of our society, these threads of the past not only provide comfort, but they are intricately woven into their present identity. Neuroscience reveals that reminiscence and acknowledgment of one's life achievements activate neural pathways associated with positive self-regard and well-being in the aging brain (Levy 2009). Achievements and the associated pride are stored in long-term memory banks, accessible through cues that can elicit both happiness and a sense of relevancy.

The act of recollecting, known as *autobiographical memory,* involves the prefrontal cortex, a region important for processing complex social emotions and self-referential thoughts. This area remains relatively well-preserved in healthy aging, enabling seniors to recall their life narratives vividly, especially when these memories are tied to personal accomplishments and milestones.

The concept of "Do you remember when we . . ." is indeed powerful. This framing anchors the listener in a shared past, and to the senior brain, it signals a moment of bonding and shared history. Such phrases engage the familiarity recognition systems, inducing comfort and feelings of social belonging, which are particularly vital to emotional well-being as we age.

The neural response to such reflections is profound. There's mounting evidence to suggest that nostalgia, the sentimental yearning for the past, activates the reward centers of the brain, like the ventral striatum. The senior brain's response to these moments of nostalgia is not just affective; it is neurochemical, with an appreciation for the recognition of life's worth being akin to other rewarding stimuli.

Recognition and acknowledgment serve as affirmations that one's actions have mattered and have left an indelible mark. These are not merely desires but basic psychological needs. For the aging population, whose societal role evolves, maintaining a sense of purpose is critical for psychological health and cognitive vigor. They seek to be remembered not just as individuals who once were, but as ones who continue to matter.

With our scientific understanding grounded, we begin to see how this translates into everyday interactions with senior audiences. Marketing that taps into sentiments of contribution and legacy can have profound impacts, creating content that echoes the internal narratives of seniors' lives and underscores their lasting value. The art of acknowledgment in marketing not only bolsters a sense of worth among seniors but also helps in fortifying brand loyalty.

Injecting lightheartedness when looking back on life's journey resonates well with the senior psyche. Such a stylistic choice in communication is a nod to the ability of seniors to laugh both with and at themselves, humanizing the brand and making any message more palatable and relatable. It's about the chuckles that come from the bygone era of big hair in the 1980s or the elder trendy ways to stay fit with jazzercise.

Taking a nostalgic trip through the decades, employing iconic cultural references contemporary to the senior's youthful years creates a bridge between past and present. Marketing could leverage this by perhaps revisiting historic sports moments, cinematic milestones, or technological breakthroughs, reigniting the emotions linked with those times.

Acknowledgment campaigns might revolve around communal or individual stories, showcasing seniors as mentors, teachers, or volunteers who continue to contribute to society in meaningful ways. By highlighting continuous involvement and relevance, organizations can foster a deepened sense of purpose and community engagement among the aged.

In crafting messages for the mature mind, one must steer from patronizing tones, focusing instead on empowerment. The narrative should not merely revere past glory but also celebrate current wisdom and ongoing contributions. With informed care, marketing and communication strategies can transform scientific insights into living tributes, ensuring that senior members not only reminisce about when they mattered but feel that they still do.

Employ GenAi to craft narratives that recall accomplishments and slipups during the long voyage of the boomer brain. This lends authenticity and connection to the boomer life.

Feeling Young at Heart – The Benjamin Button Effect

Contrary to popular belief, the aging brain does not solely dwell on the past but often maintains a dynamic view of oneself that includes qualities associated with youth. A plethora of studies affirms that individuals tend to feel younger than their actual age, a phenomenon termed *subjective age*. This discrepancy between chronological age and experienced age can affect psychological well-being and influence behaviors. For many seniors, "You're only as old as you feel" means that they are not nearly as "old" as their birth certificate says they are.

Why is this subjective feeling of youthfulness so pervasive? Neuroscience points to a multidimensional relationship between how we perceive our age and various cognitive functions. The anterior cingulate cortex and lateral prefrontal areas, associated with self-reflection and evaluation, might underlie these subjective age perceptions. When seniors consider themselves younger, it is arguably a reflection of an active, preserved cognitive process rather than a nostalgic escape.

Remarkably, believing oneself to be younger is more than a feel-good factor; it can translate to physiological benefits (Charles et al. 2019). Research has shown that those with a younger subjective age have better memory performance, less inflammation, and a lower risk of hospitalization. It's as if the brain's perception of youth can echo through the body, entraining physiological systems to a more vibrant rhythm.

A youthful self-perception corresponds with the positivity effect, wherein older individuals show a preferential focus on positive information. This bias suggests that the senior brain remains responsive to emotional stimuli and motivational goals typically associated with youth, emphasizing well-being and potential. The positivity effect reinforces a youthful mindset by encouraging a focus on life's pleasures and rewards.

Bringing this understanding into the real world, marketers can appeal to seniors by creating campaigns that mirror these internal perceptions of enduring youth. It's not about denying the authenticity of aging; rather, it's about endorsing the vitality that continues to pulse within. For instance, skincare or athletic brands that normally target younger demographics could expand their message inclusively, reflecting how older adults also value vitality and vigor.

Humor finds its place here by playfully bridging the gap between chronological age and how one feels. Lighthearted messages that poke fun at the discrepancies – such as seniors outpacing their younger counterparts in a fitness class – can be endearing and relatable. It's the sort of jocular acknowledgment that celebrates the tenacity of the young-at-heart.

Cultural representations that defy age stereotypes, from fashion to technology, can further empower seniors. Images of charismatic, tech-savvy elders enjoying the latest gadgets can bolster a youthful self-concept. It aligns with the known cognitive and emotional aptitudes of seniors, acknowledging their inclusion in the zeitgeist typically associated with younger generations.

Adventures and experiences likewise act as conduits for youthful perceptions. Travel agencies could craft compelling narratives for seniors, framed not as leisurely, passive experiences but as active explorations akin to the escapades of their youth. The language used in such narratives would seamlessly intertwine the seasoned traveler's wisdom with the spirit of adventure, captivating the imaginations of mature audiences.

At the same time, it is essential to celebrate the intersection of gained experience with continued zest for life. Content that features seniors embracing new learning opportunities, sports, or creative endeavors underscores their societal value as lifelong learners and contributors. It's about amplifying the message that the journey of personal growth and exploration is ageless. As seniors age, their desire for romance or love or feeling attractive does not diminish. If anything, it is enhanced with the newfound acceptance of oneself, and with all insecurities leaving, a stronger man or woman emerges from within the cocoon.

To weave such narratives authentically, we must anchor campaigns in the realities of senior lives, respectfully and joyfully embracing the blend of wisdom, whimsy, and willpower that defines modern aging. The stories told should not simply hark back to bygone days but must resonate with the present and future aspirations of mature minds.

Utilize GenAi in advertising campaigns to suggest new experiences, classes, or hobbies that are tailored to the senior demographic but are vibrant and youthful in spirit. This could range from digital photography classes to outdoor adventures, encouraging seniors to explore new horizons.

"Don't sweat the small stuff" – just one of the guiding takeaways that you now have from this chapter.

You know the ins and outs of marketing to this historic generation. How and why seniors respond to specific stimuli. The importance of simplicity, humor, fragrance, and more to reach and persuade this audience.

You also have the keys to the kingdom in your grasp: How to compose the most effective prompts to drive the most relevant and useful GenAi results.

Now, on to the next chapter.

PART 2

neuroAi Applications

CHAPTER 8

neuroAi Applications across Categories and Industry Sectors

You are now equipped with a body of knowledge that truly sets you apart. You have learned the inner workings of the most powerful force in your body: your nonconscious mind.

You now know how the five senses contribute so centrally to consumers' perceptions, emotional engagement, memory retention, and purchase decisions. And you know how to leverage them to maximum effect.

With this knowledge, you will be able to create the most innovative new products and the most effective marketing campaigns. You will have the ability to breathe new life into existing brands, and take them to new heights in the marketplace. You understand how and why brand extensions, digital marketing, pricing, promotions, experiential marketing, and so much more can work to your best advantage.

Such is the sheer power of neuroscience and what it teaches us. In the previous chapters we have gained extraordinary insights and learnings about how the brain works, at the deepest level – and why.

You have just read one full book about that.

But that is only *half* the epic story of our times. The second half awaits you now: the history-making advent of artificial intelligence, and the life-changing effects it is having on humankind. The applications across industry sectors is the fun part.

This book is intended not only to give you a solid grounding into what GenAi is and how it works. It is designed to arm you with the practical know-how to apply AI most effectively on behalf of your business and your career.

You will come away with a firm grasp of the seismic changes that this groundbreaking technology is bringing to our world. And you will have learned the most important lessons of all: the tools you need to know to harness AI – especially GenAi – to help you gain an unprecedented competitive advantage in the marketplace. How to make GenAi work for you.

Because make no mistake: GenAi is revolutionizing every aspect of life today, and will even more so in the future. Having the knowledge and the ability to understand AI and apply it will be the great dividing line of the twenty-first century – similar in scope and impact to the other great leaps forward in humanity's history.

What lies ahead is intended to carry you over that line and equip you for your future. We now focus on applied learnings of neuroscience in creating real products, packaging, and marketing. What principles of neuroscience must be embedded in GenAi algorithms to create products, packaging, consumer experiences, brands, fragrances, flavors, and music. These practical applications form the second half of this book.

CHAPTER 9

desireGPT – Drive Desire with neuroAi

Will they love it? Want it? Crave it? Shamelessly desire it?

Long lines in front of the Apple Store on launch day – screams desire.

Wait for the Tesla Cybertruck – shameless desire.

A Stanley coffee mug, the Nike Aerofly – raw desire.

Consumer *desire* is the dream of every marketer.

Do the endless brochures and digital product descriptions evoke desire? Does the imagery in a carousel and the accompanying language evoke desire to pause, click through, and purchase?

This chapter is about the science of desire. In it, we explore what neuroscience can tell us about the structures and the processes that are key to the formulation of desire within the brain.

It's essential information for anyone looking to connect with audiences – especially for marketers aiming to make their products irresistible. Understanding the different systems that combine to create and drive the emotion of desire will help create products and messages that will appeal directly to the nonconscious mind. Algorithms of GenAi must embed these principles of desire in the crafting of output.

The Brain and Desire – Six-Dimension Neuroscience Framework

We're on a quest to figure out what sparks desire in our minds – the neuroscientific mechanisms and how they connect to the aspects of the nonconscious we have been discussing through earlier chapters.

There are six dimensions to evoking desire that correspond to various networks and mechanisms in the brain. Activating some combination of these dimensions through carefully crafted messaging will lead to far superior connection with audiences. Let us run through these various dimensions of desire, providing a good overview. We will then consider some of these dimensions in further detail.

The first dimension of desire is built on dopamine – the concept of novelty and newness that results in reward value. Our brains assess what rewards we stand to gain from our decisions, activating regions such as the ventral striatum. What it means for marketers and GenAi algorithms: *Showing, demonstrating and highlighting the benefits of a product can directly appeal to this reward mechanism in the brain.*

This dimension is also reflected as the salience network that corresponds to the idea that what pops out at us first and stands out probably matters the most. This network involves areas like the anterior insula and dorsal anterior cingulate cortex. What it means for marketers and GenAi algorithms: *Making your message stand out is crucial in a world filled with distractions.*

Engagement with media or a product activates the "default mode network," which is active during inward focused thought. This introspective element is crucial as it deals with personal relevance, mirroring, and empathy – elements that make marketing feel like a story tailored just for the individual consumer. What it means for marketers and GenAi algorithms: *Weave stories around products that resonate on a personal level, making your audience feel like you're speaking directly to them.*

Then there is the *pleasure effect*. It is all about dopamine – a neurotransmitter closely linked to *anticipation and reward*. It courses through areas like the nucleus accumbens when we *anticipate* a reward. What it means for marketers and GenAi algorithms: *Create experiences that promise future joy. It's not just about the product; it's about the excitement it brings.*

The second dimension of desire is grounded in serenity, contentment, and the familiar, which is associated with the neurotransmitter serotonin. This neurotransmitter is often associated with mood regulation, contributing to feelings of well-being and happiness. When we bask in the glow of the familiar, the serotonin levels in our brain are likely to be elevated. This has the effect of calming anxiety and raising our sense of belonging – meaning we desire the familiar and find comfort in the predictable. What it means for marketers and GenAi: *Craft messages that convey comfort rather than issue challenges. In a world of relentless change, offering a product or service that feels like "home" can be a refuge for the weary consumer.*

We desire what is trustworthy, transparent, and honest. This is the third dimension of desire. Trust is associated with oxytocin, which is often affectionately referred to as the "cuddle hormone" or "love hormone," thanks to its role in social bonding and trust. Oxytocin facilitates the formation of trust and the development of social memory, being released during interpersonal transactions helping to augment the recognition of familiar faces and fostering trust and deeper connections. In the business context, the "oxytocin moment" becomes a linchpin for creating connections that are deeper than the superficial. What it means for marketers: *Craft an identity that is a consistent and reliable voice. This will carve a distinct identity in the mindscape and forge trust through transparency, honesty, and integrity.*

The human brain is an efficiency expert, favoring processes that are smooth and effortless. This cognitive fluency corresponds to the fourth dimension of desire – simplicity. The brain desires and prefers stimuli that it can rapidly comprehend and act on without excessive cognitive strain. What it means for marketers and GenAi: *Harness this preference for cognitive ease by embracing simplicity throughout the consumer experience, from messaging to design and decision-making.*

The fifth dimension of desire is social bonding. Neuroscientists have long contemplated the underpinnings of our tribal instincts – a need to belong that goes back deep into our evolutionary history. The core of tribalism resides in the intricate dance between neurotransmitters like oxytocin, which reinforces social bonding, and hormones such as cortisol. What it means for marketers and GenAi: *Cultivate a brand tribe by building narratives around inclusion, brand loyalty, and foster brand-centric communities that emphasize shared experiences and values.*

The sixth dimension of desire is for the aspired self. This is the highest level in Abraham Maslow's pyramid. This corresponds to self-actualization and looks for acceptance, purpose, meaning, potential, morality, virtue, creativity and spontaneity. Humans place a premium on this self-actualization and acutely desire products and services that give them that satisfaction and value. What it means for marketers and GenAi: *Embed the brand in purpose, and give meaning to the purchase of the product that is more than just the functional benefits that come with the product.*

These different dimensions all contribute to evoking desire in the minds of the consumer. Of course brands do not need to play with all of these dimensions, but will often find some of these dimensions are more relevant to their brand. With each dimension we have also seen what the key takeaway and actionable insight would be for the crafting of messaging that targets the nonconscious minds of the consumers.

Dopamine Rush: Desire from Novelty and Intrigue

Dopamine is often dubbed the "rush" neurotransmitter. It plays a central role in our brain's reward system, illuminating why we chase after new, exciting experiences. When we encounter something novel, anticipate a reward, or revel in success, our brain's dopamine circuits are activated, providing the boost we feel in these moments. This surge is the dopamine dimension at play – a fundamental aspect of human desire that underpins our pursuit of the new and exciting.

This biochemical element is what makes the launch of a revolutionary product so electrifying. Dopamine doesn't just signal pleasure; it's also crucial for enforcing learning related to rewards. The brain is hardwired to seek out situations that will instigate its release, pushing us toward moments of discovery and innovation.

Biochemically, novelty boosts dopamine transmission within the brain's reward circuitry, part of the mesolimbic pathway. When we are exposed to something new, dopamine-rich areas such as the substantia nigra and ventral tegmental area become activated. This, in turn, influences parts of the brain encircling motivation and attention, including the nucleus accumbens and the prefrontal cortex.

If you're wondering why regular patrons get so excited by limited-edition releases or why unboxing videos on YouTube garner millions of views, look no further than the dopamine dimension. The lure of the unknown and the anticipation of pleasure grips our neurological framework, commandeering our focus and engagement (Lewis 2015).

Moving from the abstract to the applied, consider the implications of the dopamine dimension for marketing. When a brand launches a new product, it's not merely showcasing features – it's activating a dopamine-driven desire for novelty. This is where storytelling comes in, setting the stage for discovery that ignites the consumer's excitement and establishes a memorable experience.

For marketers, invoking the dopamine dimension means generating a sense of freshness with every campaign. A continuous cycle of innovation and creativity is key to captivate and maintain consumer interest. It involves fostering a narrative of perpetual novelty, suggesting that with each purchase, consumers step into a new chapter of their own story.

In practice, this dimension translates to "new release" excitement. Constructing events or launches feels less like a presentation and more like a celebration of human curiosity. This spectacle is a curated ballet where every lighting cue and musical score is designed to escalate the consumer's pulse with anticipation.

But it isn't just about the big reveal. When the dopamine dimension is woven throughout the marketing campaign, it creates ancillary pathways of intrigue – a teaser here, a sneak peek there, all building toward the crescendo of release. Each touchpoint is a jigsaw piece of an enigmatic puzzle, designed to entice the brain's desire for solving the unknown.

The journey of the dopamine dimension culminates not when the product is acquired, but when it imbues awe into the tapestry of the consumer's life. The perpetual quest for the novel doesn't end; it iterates, leaving an open door for the next dopamine moment. This implies that marketers can perpetuate a cycle of desire that keeps consumers coming back for more – profoundly rooted in the brain's search for novelty.

Use GenAi to integrate all aspects of novelty, newness, unexpected pleasure, and discovery into the description of the product. The feelings to be evoked are the rush, pleasure, and excitement of a blind date.

Serotonin Satisfaction: Desire for Comfort and Familiarity

While dopamine drives our pursuit of the new, serotonin plays a comforting counterpart, nurturing our love for the familiar. This neurotransmitter is often associated with mood regulation, contributing to feelings of well-being and happiness. When it comes to desire, serotonin's influence is subtler, yet no less profound – it fosters a longing for the familiar, a comfort in the predictable.

The cerebral underpinnings of this phenomenon involve neural pathways that intertwine through several brain regions, including the raphe nuclei, where serotonin production predominantly occurs. From here, serotonin's tendrils reach out to regions like the amygdala and hippocampus, which are vital for emotional processing and memory, reinforcing the connection between familiarity, comfort, and contentment.

Of course, the love for the known is not just a matter of neurochemistry – it has evolutionary roots. Our ancestors benefited from the safety and resources that familiar environments provided. As such, our brains have been sculpted by natural selection to prefer settings and experiences that are known and there-fore predictable (Liao et al. 2011). This is the *serotonin moment*: the physiological embodiment of safety and tradition.

From a scientific vantage, serotonin's influence extends to the hippocampus, a structure deeply involved in memory. It aids in the formation of spatial and episodic memories, helping us navigate and find comfort in familiar places. And serotonin receptors dotted throughout the cortex modulate cognitive processes tied to learning and memory retrieval, subtly hinting why mom's soup tastes just like a warm embrace (Coray & Quednow 2022).

When we bask in the glow of the familiar, the serotonin levels in our brain are likely to be elevated. This has the effect of calming anxiety and raising our sense of belonging. When one reclines in a beloved armchair, there's a seroto-nergic whisper reassuring that here, in this nook, all is as it should be.

Moving into the realm of practice, understanding the serotonin moment is essential for marketers crafting a message that seeks comfort rather than challenge. In a world of relentless change, offering a product or service that feels like "home" can be a refuge for the weary consumer. This is the art of nostalgia marketing, where brands evoke cherished memories to connect on a deeper emotional level.

Marketers can weave this comfort into their narratives by emphasizing the heritage of a brand, the legacy of a product, or the enduring quality of an expe-rience. These story elements resonate with the consumer's innate desire for the recognizable, summoning serotonin's soft power to invoke a sense of contentment and satisfaction with the choice they make.

This dimension suggests that when a person returns home from an adven-turous journey, it is not merely a retreat into the mundane but a serotonin-fueled celebration of the familiar. The marketer who understands this is adept at balancing the thrill of new experiences with the reassuring embrace of the tried and true, leveraging this dynamic to keep consumers engaged.

Employ GenAi to highlight familiar and comforting aspects of the product. So the consumer feels the same delight in meeting an old friend for dinner.

Oxytocin Oasis: Desire for Trust and Validation

As already noted, oxytocin is often affectionately referred to as the cuddle hor-mone or love hormone, thanks to its role in social bonding and trust. It is a neu-ropeptide that acts as both a hormone and a neurotransmitter, influencing social behavior and emotion. When we experience social bonding, whether it's through a warm hug or a heartfelt conversation, oxytocin levels in our brain surge, reinforcing the feeling of connection.

Science has shown that oxytocin facilitates the formation of trust and the development of social memory (Shamay-Tsoory et al. 2016). During interpersonal transactions, oxytocin is released, augmenting the recognition of familiar faces and fostering the trustworthiness we attribute to them. This has profound implications for our understanding of human relationships and, by extension, consumer-brand relationships.

But oxytocin's reach goes beyond just feel-good encounters. There is ample evidence that oxytocin influences our moral compass, subtly steering us toward upholding social norms and ethical behaviors. This pivotal role makes oxytocin a linchpin in the mechanics of integrity and worthiness, two attributes highly regarded in the commercial cosmos.

The oxytocin dimension is biologically activated in moments of trust. Studies have found increased oxytocin release in response to positive social interactions, such as receiving trust or demonstrating a trustworthy act. This feedback loop of trust and oxytocin secretion enhances bonds and the sense of security within relationships, making it a powerhouse for forging loyalty.

In the business context, the oxytocin moment becomes a linchpin for creating connections that are deeper than the superficial. Successful brands often forge trust through transparency, honesty, and integrity. An endorsement by a respected figure or a genuine narrative can trigger consumers' oxytocin release, fashioning an almost palpable sense of trust and belonging.

Celebrity endorsements capitalize on the trust woven into the fabric of social familiarity, nudging the oxytocin dimension into the spotlight. The handshake between a beloved icon and a brand can release that dash of oxytocin, convincing our brains that we're not just buying a product; we're buying into a trusted circle (Osei-Frimpong et al. 2019).

Consistency in brand messaging activates a familiarity that taps into our need for stable, predictable affiliations. By crafting a consistent and reliable voice, brands not only carve a distinct identity in the mindscape of consumers but also activate the oxytocin axis, reinforcing the sense of a trusted ally in their subconscious.

Through narrative, a brand can weave a story of integrity that resonates with the oxytocin-laced desire for authenticity. Transparency isn't just a buzzword; it's a beacon that lights oxytocin's path. When a product's journey, from creation to consumption, is shared openly, it entwines product integrity with consumer trust, embedding the brand within the consumer's mental tapestry.

GenAi can craft messages that engender trust and communicate subtle authority. Spokespersons and celebrities can also be identified that further establish subtle and nuanced trust in the mind of the consumer.

Cognitive Fluency: Desire for Simplicity

Our brains love simplicity. Cognitive fluency refers to the ease with which our brain processes information, and it plays a significant role in our decision-making.

A fluently processed stimulus is one that the brain can rapidly comprehend and act upon without excessive cognitive strain. The smoother the information flow, the more positively we tend to feel about it – whether it's a product, a brand, or an entire shopping experience.

Numerous studies have highlighted cognitive fluency's role in decision-making (Jain et al. 2021). For example, psychological experiments have demonstrated that stocks with easily pronounced names outperform those with complex names, regardless of the companies' actual performance. It is as if the brain's preference for fluidity and ease seeps into our financial choices, influencing our perceived value of an investment.

Neural imaging research has provided insight into how fluency affects the brain's reward circuitry. When individuals experience fluency, there's a noted activation in regions such as the ventromedial prefrontal cortex, associated with value judgments and positive affect. This suggests that not only does fluency make for smooth cognitive sailing, but it also makes us feel good as we coast along.

Cognitive fluency doesn't just reside in the abstract realm of neural pathways; it permeates our real-world experiences. When faced with written material, for example, our brain prefers clear, concise, and familiar language. This preference is not trivial – it impacts how we interpret and trust information. Legible fonts and high-contrast color combinations are not mere design choices, they're amplifiers of fluency, increasing comprehension and retention (Lunardo et al. 2016).

The more frequently we're exposed to an item, the more fluently we can process it – a familiarity effect that makes the item seem more trustworthy and preferable (Alter & Oppenheimer 2008). Unsurprisingly then, brands seek to create familiar experiences through consistent logos, slogans, and jingles, embedding themselves into our cognitive landscape.

Learning from fluency, marketers have developed frictionless customer experiences. The fewer the stumbling blocks between a consumer and the checkout, the better. Amazon's one-click ordering is a quintessential example – reducing the purchase process to a simple, unambiguous action has transformed online shopping.

Marriages of form and function, like that of Apple's user interfaces, are testaments to fluency in action. Intuitive gestures and minimalist design further streamline user interactivity, courting the brain's fluency bias. Apple's success well illustrates how ease of use and aesthetic simplicity can become core components of a brand's identity – and a source of competitive advantage.

Too many features or options can lead to decision paralysis, an unintended anti-fluency. Market leaders harness this aspect of cognitive ease by offering curated sets of options or guiding consumers through decision-making with clear, step-by-step processes. Simplicity isn't just a design principle; it's a form of consumer respect (Adriatico et al. 2022).

Use GenAi to analyze and simplify complex information into clear, concise, and easily digestible content. This can involve rephrasing technical jargon, creating informative infographics, or breaking down complex concepts into simpler parts, making the information more accessible to a broader audience.

Tribal at the Core: Desire for Deep Belonging

Humans are inherently social creatures with a deep-seated need to belong. This drive for inclusion and community is rooted in our evolutionary past and is reflected in the way our brains respond to social stimuli. Neuroscientists have long contemplated the underpinnings of our tribal instincts – a need to belong that stems from early human societies, where survival hinged on communal living. Functional magnetic resonance imaging (fMRI) studies reveal that social exclusion activates areas in the brain like the dorsal anterior cingulate cortex (dACC), demonstrating the pain of ostracism (Eisenberger et al. 2003).

The core of tribalism resides in the intricate dance between neurotransmitters like oxytocin, which reinforces social bonding, and hormones such as cortisol, which reflects the stress of social disconnection. Oxytocin, in particular, facilitates the formation and maintenance of social bonds and feelings of trust, crucial for a sense of community.

This yearning for connection goes beyond just company; it taps into recognition and shared identity (Xu et al., 2019). The mirror neuron system offers insights into how we emulate and learn from others within our "tribe," shaping our behaviors and aligning them with community norms. Being part of a group means sharing gestures, language, and emotions, which are all reflected in this sophisticated neural circuitry.

Diving deeper into our cerebral programming, we encounter the lateral septum, an area implicated in bonding and group behavior. Studies in rodents have shown that neurons in the lateral septum are activated during social interactions (Menon et al., 2022). In humans, this may translate to the warm buzz of a tight-knit gathering, or the synchronized claps and cheers in a crowd.

A marketer's challenge lies in cultivating a brand tribe. Just as sports fans wear their team colors with pride, consumers can become champions of a brand, echoing the essential tribal call of "us vs. them." Marketers can harness this by creating brand-centric communities that emphasize shared values and experiences.

Initiatives like loyalty programs or exclusive memberships are more than mere sales tactics; they are offerings to belong to an inner circle, tapping into the deep-rooted desire for affiliation. A limited-edition sneaker drop isn't just a product launch; it's initiation into an exclusive club. Such strategies can create an "oxytocin rush" that cements loyalty and advocacy.

To imbue campaigns with this sense of camaraderie, marketers can play on in-group symbols and languages that resonate with the brand's community. Employing social media as a digital campfire, they can curate spaces where consumers share stories, jokes, and experiences, further strengthening the neural ties that bind the tribe.

The journey toward constructing a community within commerce is not without its challenges. Yet, by nurturing a narrative that aligns with consumers' deep-seated social drives, brands can become more than entities. They can become bastions of belonging where every transaction is underpinned by a powerful sense of *we*.

Employ GenAi to generate stories and narratives that reflect the shared experiences, aspirations, or challenges of the community. These narratives can foster a strong emotional connection, making buyers feel part of something bigger.

Self-Actualization: Desire for the Highest in Me

Self-actualization is the highest level in Maslow's hierarchy of needs. The layers of Maslow's pyramid represent the value humans place on their needs. Contrary to popular belief, needs for self-actualization trump esteem, love and belonging, safety, and even physiological needs.

Physiological needs include the simple human needs for clean air, nutritious food, unpolluted water, protective shelter, good sleep, clean clothing, and the opportunity to procreate. Safety needs generally center on personal security, stable employment, having a home, having income and savings, feeling safe and healthy. The need for love and belonging transfers to the emotional domain and represents the needs for different kinds of connection ranging from friendship, intimacy, nurturance, belonging, bonding, connection and support. Esteem needs pertain to identity and respect and reflect expressed recognition, status, strength, respect, achievement, boldness, uniqueness, strength, confidence, and freedom. While all these needs seem important, they remain subservient to the needs for self-actualization – to truly aspire to be the best one can be. This need is aspirational, and represents the evolved self of the human being. Virtue, morality, spontaneity, experience, acceptance, realization, meaning, and purpose color this dimension and represent the highest of human yearnings.

GenAi algorithms that structure messaging and innovation to align with Maslow's self-actualization layer will find rich rewards. It is at this layer that customers become price insensitive, and are drawn to the better version of themselves. Sustainability, purpose-driven missions, companies, and products let consumers find the better version of themselves in the brand and the product. Judicious use of the aspired self in the consumer stimulates desire.

In this chapter we have discussed the six-dimension neuroscience framework that underpins desire. We have laid out how GenAi algorithms can intelligently use this framework to both evaluate the amount of desire created by marketing and messaging. The same framework can be used by GenAi algorithms to create desire in the minds of consumers as well.

CHAPTER 10

The Consumer Experience

T he brain loves compelling experiences.

Experiences that were incredible. Memorable. Awesome. Just an utter pleasure.

Remember that old saying, "It is not what they said, but how they made you feel." It is all about how we felt about that experience.

But what then is a "great experience"? Is there a way the brain defines it?

In this chapter we parse what a great experience is and how to use GenAi to create it for consumers at every touchpoint. Whether it is a humble chatbot, a complex kiosk interaction, or a healthcare condition management application trying to create a superior interactive experience, the foundational principles are the same.

It is important that we teach algorithms how to create superior experiences that appeal to us. The experiences create habits, and create compelling desire to experience it again. Habit creators, chatbot makers, and human-machine interaction creators, take note.

Information – What Matters Most?

In a world saturated with information, deciphering the essence of a product or an experience seems like a monumental task. For marketers, crafting a compelling narrative has long been the key to captivating hearts and minds. Yet, a singular metric often steals the show. Neuroscience explains the human brain's desire for simplification – our cognitive tendency to reduce complex data to its most fundamental components for easier processing. This preference for simplicity is evident in the psychological concept known as the *cognitive miser* (Dunn & Risko 2019), which suggests that our minds conserve energy by favoring mental shortcuts over elaborate thinking processes.

When presented with multifaceted information about products or services, consumers instinctively seek a single number that promises to dictate the superiority of one choice over another. Notably, this search for a singular metric aligns with our heuristic biases, where a single attribute can disproportionately influence our judgment, also known as the *anchoring effect* (Furnham et al. 2011).

Such singular metrics serve as beacons in the bewildering sea of choices. Consider the example of cameras, where megapixels once reigned supreme as the numerical torchbearer of quality. Tech products parade their gigahertz and cores, while the beverage industry touts calorie counts to appeal to the health-conscious, each digit wielding a psychological influence that extends beyond its numerical value. We call this a *category busting metric*. Be it calories in a cola, amount of cacao in chocolate, the number of camera lenses on an iphone, the gigahertz of computer, the horsepower in a car, the single number becomes the primary differentiator.

Despite the appeal of singular metrics, neuroscientific research suggests that overreliance on one number can lead to what is known as *attribute fixation*, where the diversity and synergy of other qualities are overshadowed. Balanced communication that incorporates a singular metric while still conveying the richness of the product's features can ensure a more holistic consumer understanding and avoid cognitive oversimplification.

The ideal approach balances the delivery of this information amidst a narrative that sustains the consumer's emotional and cognitive engagement. Apple's adroit marketing of the iPhone's camera lenses demonstrates this by not only focusing on the number of lenses but also conjuring a story of innovation and versatility.

In practice, embedding the allure of singular metrics within a broader narrative demands an understanding of the neurobiological underpinnings of storytelling. The human brain's neural pathways light up in response to stories, which provide a scaffold for memory and emotional resonance (Mar 2011). Marketers must craft campaigns that merge these narrative arcs with the might of a single, potent metric, ensuring the number is not an abrupt interruption but a harmonious note in the melody of the message.

Utilize GenAi to determine the metric and sets of metrics that matters most to the consumer. Identify the ways and means of describing that metric to the consumer.

Interaction – Try before You Buy

Humanity's evolutionary success owes much to our ability to engage dynamically with our environments. This interaction isn't merely physical; it's deeply ingrained in our neural circuitry. Modern neuroscientific research aligns with the theory of embodied cognition, which posits that our understanding of the world is fundamentally shaped by our bodily interactions. The brain's motor regions, such as the prefrontal cortex, light up not only when we execute movements but also when we anticipate or imagine them (Liang et al. 2002). The implications for engagement with products and services are profound. Tangible interaction with an object can heighten the mental imprint it leaves, potentially boosting the perceived quality of an experience.

The sensory integration theory explains how the brain combines information from various senses to form a coherent and comprehensive understanding of one's environment. In real terms for the consumer, this translates to a multisensory encounter with a product potentially being more memorable and satisfying than a unisensory engagement. When customers can touch, see, and interact with a product, their experience is enriched, leading to a more substantial cognitive and emotional reaction.

The relationship between haptic sensations and memory formation is robust, with studies showing that manual engagement can improve memory recall (Streicher & Estes 2015). When consumers physically interact with products – such as testing a gadget or flipping through the pages of a book – the manual activity can enhance recall and solidify brand recognition. Marketing strategies that encourage hands-on interaction can capitalize on this link between tactile experience and memory reinforcement.

The process of decision-making involves complex neural networks that include systems for value computation and prediction error signaling. Interactivity can amplify the perceived value of an offering through enhanced engagement and can reduce the prediction error by providing direct experiential data. By strategically integrating interactive elements into marketing, companies can influence the decision-making process, nudging consumers toward a favorable outcome.

We crave stories that we can step into, ones that we can interact with and not just passively consume. Reality shows and improv comedy thrive on this interactive storytelling – they live in the space where narrative meets participation. This approach in the context of marketing creates an immersive tale that the consumer takes an active role in, merging data with heightened engagement. Think iPhones and their commercial stories of shared photos and memories, inviting viewers to picture themselves within the narrative, tapping into the allure of the interactive tale.

Interactive marketing doesn't simply sell; it tells a human story, a tale where consumers are not mere spectators but participants in a shared experience. It is in this mutual experience where the heart of marketing beats strongest. From the open invitation to touch an Apple device to the personal engagement one feels in an IKEA showroom, these moments of interaction aren't just marketing techniques – they're connections that humanize the brand and elevate the exchange from transaction to interaction.

Utilize GenAi to create interactions for the consumer that allows them to experience the product or the service in meaningful ways. Allow the consumer the joy of use, the joy of discovery, and facilitate a way for the consumer to sell themselves the product or service.

Entertainment – Emotional Relief from the Process

The human brain not only seeks novelty but revels in it. Novelty acts as a cognitive stimulant, provoking our interest and encouraging a state of heightened awareness. When something is entertaining, it often contains elements of novelty, surprise, or humor that activates our brain's reward centers, namely the mesolimbic pathway responsible for dopamine release. This reward reaction not only encourages continued attention but creates stronger memories associated with the entertaining content.

Emotional experiences tend to be more vivid and more easily recalled than those that are emotionally neutral. Known as the *affective salience hypothesis,* this concept suggests that emotionally engaging content, such as entertainment, can enrich the encoding of memories and facilitate their retrieval (Cooper et al. 2019). This is particularly important in contexts where information retention is crucial, such as in educational settings or in conveying complex services like financial advice. Emotional relief creates a "cognitive relief" from the cognitive distress in the assimilation of facts, and rejuvenates the brain to come back and learn more. Great professors in academia do this regularly. Professor Richard Feynman was adept at this, and his lectures have such broad appeal because of his extraordinary capacity to make physics entertaining.

Beyond leisure, entertainment can have a significant impact on learning. The idea of "edutainment" – a fusion of education and entertainment – highlights the cognitive benefits of combining enjoyment with information acquisition. An entertaining approach can facilitate learner engagement and knowledge retention, and it can dramatically improve the assimilation of complex information such as financial concepts or healthcare instructions that might otherwise be regarded as austere or prohibitive.

Stories can be inherently entertaining, possessing the unique power to captivate and persuade. The narrative transportation theory explains how engrossing narratives can lead individuals to experience deep emotional connections and shifts in attitudes and beliefs. In sectors like finance or healthcare – where engaging the consumer takes on paramount importance – the integration of narrative can transform sterile information into enthralling content, making it more palatable and impactful.

In the financial world, terms and data can be dense and esoteric. Entertainment introduces a human element, a way to simplify and personify complex information. When banks and financial advisers harness storytelling and humor, they mitigate the intimidation factor, making financial planning more approachable and engaging, which is essential for building trust in long-term relationships with clients.

Insurance, often a grudging necessity, can be transformed through entertainment. By creating engaging content that taps into relatable situations, humor, and the universal appeal of narratives, insurance companies can foster closer bonds with policyholders – shifting the perception from a faceless contract to a personal and comforting safeguard.

In conclusion, whether it's the parables in sacred texts, the comedic brilliance of a Feynman lecture, or the infectious joy of a Disney theme park, entertainment has consistently proven to be a powerful mediator of information and a catalyst for emotional and cognitive engagement. Businesses traditionally veiled in sobriety can harness this power to elevate their consumer experience, engaging the brain in a way that bullet points never could. Through entertainment, serious subjects become relatable, personal, and memorable, transforming the monochromatic to the multicolored, and enriching the dialogue between business and consumer.

Use GenAi to create and provide entertainment to the consumer that is brand appropriate, product specific, and context tuned. This allows for a rich and vibrant consumer experience.

Education – Little Bytes of Wisdom, and Tasty Nuggets of Knowledge

Although the human brain is naturally curious, there are certain preferences for the way it receives information. Neuroscience suggests that, in contrast to the usual academic deluge, the brain enjoys "knowledge snacks" – tiny bits of information that offer fleeting insights. When it is straightforward but insightful, and it feeds the cognitive self with a sense of creativity and enlightenment, this educational sustenance is easily consumed. These tiny but flavorful nuggets of information strengthen the structure of our cognitive processes in addition to being delicious to the mind.

Studies into working memory, such as those by Miller, suggest that humans engage more effectively with information when it's segmented into digestible parts, permitting our neural faculties to process and assimilate details without being overwhelmed. These knowledge snacks invigorate our neural appetite and inspire further intellectual satiation, reinforcing our brain's capacity for learning in an enjoyable and efficient fashion.

The innate human trait of curiosity is a driving force in the pursuit of learning. Neurologically, our brains demonstrate a preference for the joy of discovering something new. This dopamine-mediated response, as elucidated in the works of Schultz, underscores our preference for educational content that is not just nourishing but also novel and captivating.

Infusing consumer education with engaging, palatable bites of information can generate positive emotional responses and, in turn, reinforce memory and understanding. Engaging facts – akin to adding the perfect spice to a meal – stimulates the learning experience, making each new piece of knowledge not only a lesson but a pleasure. It is through this garnishing of education with appealing content that we nurture a genuine hunger for learning.

Knowledge snacks delivered through compelling stories or intriguing questions stand a greater chance of resonating with learners. Narrative transportation theory supports this, showing that engaging stories enhance receptiveness and learning effectiveness (Moore et al. 2020).

Translating the science of small-scale learning into practical applications requires creativity and awareness of the learner's mental palate. Marketing education can provide consumers with knowledge snacks that are not only factual but also provocative, stimulating a desire to dig deeper. By artfully serving up these savory science bites, we invite the audience to savor each learning moment and to delight in the taste of newfound knowledge.

Utilize GenAi to unearth small knowledge bits that can be provided to the consumer at every interaction. This stimulates playfulness, curiosity, wisdom, and implicit authority.

Simplicity – The Rule of Threes

Simplicity in the marketplace is a concept that is highly valued for its strength and, at the same time, for its subtlety: simplicity. The essence of simplicity lies in cognitive fluency, or our ability to digest information with ease. What makes an experience simple or complex? The question of how easily the brain can navigate a notion or ritual has been answered by science. At its best, simplicity resonates with the brain circuits that yearn for comfort and comprehending clarity.

Decades of psychological research uphold that cognitive load – the amount of mental processing power required to use a system – inversely correlates with simplicity and ease of engagement. An overture to this idea is the rule of threes, a notion that encases simplicity in an elegant triptych. This rule presupposes that a triad, a collection of three elements, has an innate attraction that is unmistakably appealing to the human brain. The rule reflects a pattern observed across time, culture, and context, pointing toward a fundamental neurological preference.

The magic of the number three extends to various cultural and spiritual realms, hinting at a universal structural preference. Within Christianity, there is the Holy Trinity, and Hinduism speaks to Brahma, Vishnu, and Shiva. This tricolon compels our cognitive processes and fosters a belief that tasks encompassed by three units are inherently more doable and less intimidating. The power of such a simple structure is it supplies both a cognitive and temporal scaffold that our brains anchor to with ease.

Simplicity through the lens of the triad does not merely connote brevity but also coherence and order. It aligns with the limits of working memory, particularly George A. Miller's proposition that the average number of objects an individual can hold in working memory is about seven, plus or minus two, bringing about ease in the retention of information. Reducing cognitive overhead through digestible chunks allows for more efficient mental processing and recall.

By delineating a world often overwhelmed by complexity into smaller, more manageable subparts, a triadic framework helps people manage and retain information more effectively. This is particularly relevant in a marketing environment, where consumers are bombarded with choices and messages. By refining communication to a trio of bullet points or steps, marketers make their pitch more accessible and the mental uptake of their audience sharper. For instance, the archetype of the "three-step process" is an intuitive simplification that can transform an intricate product like a mortgage into a mental canvas of simplicity.

Science informs us that the human brain is an impressive energy-consuming organ, using up about 20% of the body's energy despite being only 2% of the body weight. This high metabolic demand underpins why our cognitive system is designed to conserve energy wherever possible by gravitating toward simplicity and ease of processing. This principle plays a pivotal role in the marketplace: Simple messages are the ones that resonate because they demand less neural effort to decode and appreciate.

Elaborating on the scientific underpinnings, the transition into practical marketing application begins with the elucidation of product features in three clear, salient points. The artful distillation of information into a triad harnesses the psychospiritual allure that enraptures our attention. Humor, too, can be a powerful ally in this realm; presenting product advantages in three humorous vignettes engages the consumer emotionally, enhancing recall and favorability.

The savvy marketer crafts narratives that adhere to the rhythm of simplicity. Apple's iconic "Thousands of songs in your pocket" for the iPod encapsulated a world of functionality in a single phrase, adhering to the precepts of simplicity while evoking a grand narrative. In a similar vein, any complex service or product proposition, when neatly encapsulated in a triadic statement, has the power to become enchanting in its uncomplicated nature.

Use GenAi to simplify processes to three steps, claims to three words, messages to phrase of three words, and create the perception of ease and simplicity in the mind of the consumer.

Self-Worth – I Am Wonderful, Am I Not?

Embedded within the human psyche is the perennial quest for self-affirmation. Self-worth, a construct existential in nature, defies mere vanity, sculpting the contours of identity and personal value. Neuroscientifically speaking, self-worth has been tied to the intricate operations of the prefrontal cortex (PFC), where self-reflection and evaluation are seated. The PFC enables us to ruminate on personal narratives that configure our self-esteem.

Oxytocin, often heralded as the "love hormone," has been implicated in generating feelings of self-worth and social bonding (Kosfeld et al. 2005). When extended into the realm of consumer experience, the deliberate nurturing of a customer's self-worth becomes a critical variable. As a sales associate heralds a patron's fashionable choice, oxytocin release is potentially fostered, cultivating a sense of belonging and validation.

Furthermore, mirror neurons, which fire both during action execution and observation, play a subtle yet potent role in the theater of flattery. The cordial exchange in retail – mirrored smiles and enthusiastic affirmations – triggers the neural circuitry associated with imitation, which may deepen the customer's emotional resonance with the brand.

It is within this intimate interplay of neurobiology and social encounters that the marketer plants seeds of customer loyalty. For it is not the mere reflexive flattery dispensed with abandon, but the calibrated act of affirming a customer's choice that weaves a narrative of empowerment and self-enhancement (Shin et al. 2022).

The try-on experience in a clothing store is where customers often come face to face, quite literally, with their sense of self-worth. The dressing room transitions into a stage where a customer's self-dialogue is mirrored, echoing, "You are worth it. You are the paradigm of grace." The application starts with psychology-backed practices – team training that emphasizes authentic compliments over rote service scripts can result in a warmer, more personalized customer experience. Bringing levity to the interaction by complimenting a client's style with humor, "That tie could lead a boardroom revolution!" can make the praise more memorable and impactful.

Escalating self-worth works beyond affirming words; it delves into crafting experiences worth mirroring – productively utilizing down-time. An elevator with mirrors is no mere coincidence but a calculated engagement tactic, nudging customers toward pleasant contemplation that they, and by association the brand they engage with, are remarkable. Present your product not just as a commodity but as a companion to the customer's storyline. Just as a well-timed joke lightens a speech, humor, when tastefully integrated, can enhance a customer's self-concept. A car ad emphasizing the GPS feature might quip, "For the explorer in you who only gets lost on purpose."

Each element of marketing cultivates a customer experience that is infused with personal value. From the ambiance of a store that radiates warmth to loyalty programs that spell out exclusive benefits in three simple steps, marketers animate a world where each customer is not just another number but a valued player in their boutique production.

GenAi can identify messaging and opportunities to boost the consumer's self-worth at every interaction. This positive feedback creates habit and desirable addictions to continue to persevere in the consumer journey. Instead of GenAi showing how smart "it is," the focus should shift to showing how smart the consumer is – boost self-esteem to win.

Community – Sense of Belonging

Marketers can leverage human biology by appealing to the deeply rooted social instincts of the community. Individuals are often attracted to brands or products that enhance their sense of community and belonging (Muniz & O'Guinn, 2001). This is where the clever marketer can transform the "ME" into "WE," fostering a sense of inclusivity and shared identity around a product or brand. It is a subtle but powerful shift designed to align with the brain's neural networks associated with social connection (Baumeister & Leary 2017).

In true synchronization with the brain's chemistry, marketers strive to create experiences that offer a singular metric of connectivity. This can manifest as a car buyer who not only purchases a vehicle but also joins a car club, intertwining personal identity with communal affiliation (McAlexander et al. 2002). It's the neural equivalent of finding one's tribe – the invoking of our hardwired social nature to foster brand loyalty and customer retention.

But how do we translate this science into application in a way that resonates with the marketer? It is a blend of precision and creativity, akin to crafting a cocktail that awakens every taste bud while soothing the spirit.

Consider the inclusion of visuals in marketing materials. As trivial as it may seem, a picture of individuals actively engaging with a product or experience serves to harness the community effect (Escalas & Bettman, 2005). The group photo at the end of a corporate brainstorming session isn't just a memory – it's a snapshot of social belonging, a neuron's whisper saying, "You're part of something greater."

A car club, Sam's Club card, a Costco membership, and the Centurion card of American Express all repeatedly reiterate the privilege and joy of belonging.

GenAi must create messaging and experiences that showcase that sense of belonging that is a critical part of a superior consumer experience.

Memorabilia and the Mind

The human brain is an astonishing collector of souvenirs – each memento carrying a narrative weight that tethers to a memory or an emotion. Beyond the anecdotal, the hippocampus and the amygdala work in concert to encode and retrieve memories, particularly those with emotional undertones. When an experience is unique and emotionally charged, the brain is more likely to record it as a significant memory, reinforcing it further if there is an accompanying physical token.

The psychological phenomena of the *endowment effect* holds that people ascribe more value to things merely because they own them, suggesting a deep personal connection with our belongings. Memorabilia are powerful because they don't just represent ownership but resonate with personal identity and self-reflection, becoming part of the narrative of who we are.

The attachment we form with objects as representations of experiences is associated with the brain's default mode network (DMN), a series of connected brain regions that are active when we're engaging in introspection and remembering the past. When we hold a token of memorabilia, it triggers this network, enabling a mental time travel of sorts, revisiting the scenes associated with it.

Viewing personal photographs, or even selfies, can activate the prefrontal cortex, which is implicated in self-referential processing and personal relevance. Thus, a photograph taken at the time of a significant event – a practice not uncommon at car dealerships as mentioned – can become a cherished keepsake, triggering a recounting of our experience every time we glance at it.

The marketer, acknowledging the power of personal mementos, can craft experiences where memorabilia are not just handed out, but made personal. An olfactory trademark scent may mark the ambiance of a hotel lobby. Months later, the scent, encapsulated in a candle given as a guest parting gift, can rekindle reflections of a delightful stay. Here, the tangible meets the conceptual in a symphony of memory-induced customer loyalty.

The science of memorabilia is met with the art of personalization. In boutique chocolate stores that invite customers to craft their own blend, the resulting confection is imbued with the signature of the creator's taste – more than a token, it becomes a narrated piece of personal history. This harmony of experience and personal effort leads to memorabilia that carry exceptional sentimental weight.

In marketing, the careful curating of memorabilia is an act of engaging customers beyond the transactional. By providing consumers with a photograph, encapsulated scent, or a personalized culinary creation, they're handed not just products but pieces of a story they're eager to retell.

Employ GenAi to identify memorabilia that incorporate the highpoints of an experience in a way that resonates with the audience. Memorabilia crafted well recall a signature moment in a manner that's not easily forgotten.

In this chapter, we have captured the dimensions of what constitutes a superior consumer experience at the consumer touchpoint. These elements are to be viewed as items in a doctor's prescription, and not as items in an a-la carte menu. That is to say, an experience becomes superior if *all* elements are incorporated into the experience.

CHAPTER 11

brandGPT – neuroAi-Powered Design of Brands

B rands.

We love them. We stand in long lines on launch day.

We are loyal to them.

What attracts us to certain brands, and not others? Why will we form lifelong allegiances to some and ignore competitors?

The answers to these and many other questions about brands lie in the nonconscious. Modern neuroscience has cracked the brand code.

In this chapter you will learn many of those answers. You will also learn how to develop and deploy the most effective packaging, promotions, and other aspects of brand marketing. You will learn the secrets to successful brand extensions. And simultaneously, you will discover the sheer power of GenAi, and how to leverage it to drive success for your brand.

The brain is not merely a passive information processor when it comes to brand comprehension; rather, it is a dynamic environment where memory and perception combine to create a stage for brand resonance. Brands are mapped not just by their visual appearance but also by the emotional and cognitive resonances they elicit in the nonconscious mind. To dive deeper, we must explore the dimensions through which the brain perceives brands, starting with the larger memory structures and progressing toward the specific activations.

Face of the Brand: Visual Identity – The Form Factor

Imagine a symphony of images and colors, each with its own tone and melody, converging to create a unique visual identity. This is the orchestra in our minds that plays when we see a brand's shape: logos, colors, typefaces, and the general visual design. Each part is a separate instrument that helps us recognize the brand right away. The brain's preference for processing visual information quickly and efficiently underscores the importance of a brand's form factor in making lasting impressions.

The preference for visual economy is not random; it comes from the need to be as efficient as possible over time. The fusiform gyrus, a brain region specialized for facial recognition, is a testament to this neural frugality (Kosyakovsky 2021).

Cognitive ease, the fluidity with which our brains process information, plays a pivotal role in the form factor's appeal. A distinct logo or a unique color palette minimizes cognitive load, facilitating a feeling of familiarity and preference – characteristics highly valued in the heuristic-based decisions consumers often make.

The visual identity of a brand acts as a beacon that can create long-term memory structures within the hippocampus, the brain's hub of long-term memory consolidation. Through repetition and exposure, the form factor becomes entwined with a brand's semantic identity – its meaning-laden embodiment in the consumer's mind.

Now that our neuroscientific canvas is exposed, the creative marketer needs to become an expert in color contrasts and visual design to create a brand's iconic hallmark. Consider the iconic swoosh of Nike or the golden arches of McDonald's – such brands are in the public's neural fabric through signature motifs that are easily recognized.

These are not mere artistic choices; they are strategic brand elements informed by neuro-aesthetics and are designed to evoke a seamless recognition response, enabling positive emotional associations – warmth for the golden arches, perhaps, or dynamism for the swoosh (Chatterjee & Vartanian 2014). Color is intimately connected to mood and emotion. So picking and choosing colors that connect to the emotions the brand aspires to evoke is important – it is not just pretty, but communicates affective meaning. Similarly, lighting and dynamic range communicate more meaning than just illumination. Beams of light cutting through the fog stand more for hope and clarity than for just visual theatrics. Contrasts and backgrounds provide context. The same plain white teacup on a gold table in a quiet room stands for understated luxury, while in the cluttered environment of a coffee shop it signals the daily grind of the common man or woman. This meaning and understanding must be embedded in GenAi algorithms that create brand identities.

We may personify logos as the endearing faces of brand identities, which we could instantly identify from across a busy supermarket as if we were seeing an old friend. This anthropomorphic angle isn't just amusing – it is a bridge toward emotional engagement. Symbols of the brand deeply communicate the semiotic identity of the brand, and our brains love symbols, especially the greater meanings they might convey.

Finally, a note on font structures: They, like the human face, are particularly important as they deliver on both the aesthetic and the functional. They must be deeply linked to brand semiotics with clear and deep connections. Whether the font flows with flourishes, or is solid and brutal, or evokes typewritten nostalgia of another day, fonts speak louder than words.

Use GenAi to generate or refine brand imagery, logos, and symbols that are simple, distinctive, and easily recognizable. These iconic visuals become the shorthand for the brand in the consumer's mind, facilitating quicker recall and recognition.

Apply GenAi to analyze and select a color palette that not only aligns with the brand's identity but also evokes the desired emotional response. Colors play a significant role in visual economy and brand recall, as they can be easily associated with specific feelings or brand traits.

Sealing the Neural Deal: Semantic Identity

Delving into semantic identity, it is not just about the brand being visually distinct but also about being semantically potent. A name like Google, a neologism, stands out not only visually but also linguistically in the cortex devoted to language processing. Such a distinctive semantic footprint bestows upon the brand a cognitive spotlight, making it a directional vector in consumer choices.

Semantic identity is deeply connected to brand semiotics. The archetype of the brand permits and disallows the use of concepts, words, and phrases that connect deeply to the archetype of the brand. This is a perfect use case for GenAi – exploring the semantic identity of the brand through the symbols, words, contexts, and imagery that can and should be associated with the brand. It is important to blend into the semantic identity, the understanding of the brand archetype. The brand archetype connects deeply to the self-image of the consumer, and the consumer archetype – the hero within that the brand enables connection to.

Translating these principles into marketing strategies, one might opt for a brand name or tagline that is not only unique but rolls off the tongue with ease, sticking in memory like a catchy tune (Stocchi et al. 2016). The linguistic tool can render the brand a habitual reference within the consumer's dialogue, both internal and shared.

Utilize GenAi to generate memorable slogans or taglines that encapsulate the brand's essence. Well-crafted slogans are a powerful tool for reinforcing brand memory structures through semantic identity.

The Indispensable Function: Not One of the Many, but the ONLY

The dimension for any successful brand is its indispensable function – the fundamental service or characteristic that sets it apart in the consumer's brain. This is no mere characteristic; this is the signature tune that resonates across the neural network of your target market. Consider the smartphone: not primarily a phone anymore but a gateway to photography, communication, and endless entertainment. We begin our scientific exploration by delving deep into the concept of functional fixedness in cognitive psychology, a principle that illuminates how the mind becomes attuned to specific uses of an object – composing the function as not only being indispensable, but the brand as the ONLY provider of the function sets it apart.

Items are seen according to their most frequent usage, according to the theory of functional fixedness. Breaking free from this psychological barrier leads to innovation (McCaffrey 2012) and rebranding – an essential movement for any brand seeking to highlight its indispensable feature. Innovations in smartphones, for instance, have overhauled the main function of a phone with features like high-quality cameras, which now outshine the original consumer value.

The neural disposition toward prioritizing certain features over others is prioritized by the reward system in the brain. Dopamine, a neurotransmitter associated with reward and pleasure, plays a vital role when consumers engage with these indispensable functions. The satisfying click of a phone camera and the stream of notifications are prime examples of such rewarding experiences that inscribe brand desirability into the neural circuitry.

Transitioning into application, brands must leverage these insights to sculpt their identity around the feature that elicits this pleasurable dopamine response. Doing so fortifies the brand's position in the consumer's neural framework, becoming the "go-to" feature synonymous with the brand. Thus, the marketer can shift focus, transforming a common item into something extraordinary that promises a neurochemical reward.

One way of achieving this is by creating engaging sensory experiences around a brand's indispensable function. These experiences can alter neural connections and can lead to a rebranding even within the confines of functional fixedness. For example, the haptic feedback on a smartphone when a photo is taken instantaneously gratifies and establishes a tangible anchor for the brand's identity (Huntone 2016).

Neuroplasticity, the brain's ability to form and reorganize synaptic connections, particularly in response to learning or experience, allows for flexibility in consumer perceptions of a brand. The conscientious marketer can guide this process through targeted marketing campaigns that repeatedly associate a brand with its indispensable function, effectively creating pathways that light up at the mention of the brand.

In this journey from neuromarketing perspective toward practical application, lighthearted storytelling linked with the brand's key feature can significantly impact consumer recognition. A humorous ad highlighting the absurdity of a world without the brand's indispensable feature can be effective in reinforcing this mental connection.

Ultimately, in an increasingly saturated market, the indispensable feature becomes the brand's unique signature. It's the standout that resonates with clarity and distinction, ensuring that consumers will always tune into the brand that truly understands and delivers what is most important to them. Smart marketers will, therefore, compose brand narratives and experiences that consistently amplify this note, ensuring it is heard over the din of competition.

In short, marketers need to make sure that their brand's essential feature is not only visible but also the main point of the experience for the customer. Like a master who knows how powerful the ending of a symphony can be, a marketer must find and arrange the part of a brand that customers will think is the best thing ever.

Employ GenAi to conduct market and competitor analysis, identifying gaps or areas where the brand excels. Highlighting these unique selling propositions (USPs) in advertising campaigns helps to solidify the brand's position and essential features in the consumer's mind.

Feelings That Rise Spontaneously at the Mere Mention of the Brand

Emotions are the brain's quiet language. They tell deep stories that have a bigger impact on decision-making than we usually know. Feelings about a brand are very interesting, like how a long-lost perfume can make you remember and feel and change how you act. Neuroscience has taught us a lot about how feelings and thoughts work together to shape brand choices.

It is a big stage in the brain where feelings are both viewers and players in the story of buying. For example, the amygdala is a powerhouse for processing emotions. It responds to events with emotional energy long before we become aware of them. This way of thinking about the brain explains why some brands can make you feel happy or trusting right away.

From an evolutionary perspective, emotions served as a rapid response system, ensuring swift action to stimuli (Moody et al. 2007). Today, this manifests in brand interactions as gut feelings – those instinctive reactions that guide our choices. A brand can evoke a spectrum of emotions from security to pleasure, which become inextricably linked to the consumer's neural representation of that brand.

The limbic system acts in concert with cognitive processes, particularly the orbitofrontal cortex, to evaluate the reward value of a brand's feelings. A sense of discovery at an Apple store may trigger a novel neural choreography, merging the brand experience with a sense of innovation and sophistication that consumers seek.

Sentiments must be sincere and consistent to solidify this emotional bond. When the value proposition of affordable luxury is echoed in every detail, from storefront to shopping bag, it anchors the brand sentiment within the consumer's emotional spectrum.

To summarize, a marketer must become the choreographer of emotions, crafting a suite of feelings that entice and retain customers. It's not merely about selling a product; it's about composing an emotional escapade where the consumer is the hero, and the brand plays a leading role in their journey of discovery, joy, and fulfillment.

Use GenAi to identify feelings naturally associated with the brand. Use GenAi to create content that evokes the feelings deeply rooted and connected to the brand.

Leverage GenAi to develop compelling brand narratives that highlight the emotions evoked by the brand. Nike does not sell shoes, but sells the sweat of accomplishment – the trials, tears, and eventual triumph of accomplishment.

Branding beyond Products: Values-Based Networks and Communities

Beyond the physical shapes of goods and services, branding values are the moral and social signs that connect a brand to something bigger and more universal. We can see how strongly these values are rooted in our decisions and sense of self thanks to neuroscience. Supporting a brand's values is a deep connection between our personal identity and the brand's attitude. This is where a powerful form of brand loyalty lies.

The concept of implicit egoism suggests that people are drawn to things that they nonconsciously associate with themselves. When a brand vocalizes values that mirror our own, it cements itself within the narrative of who we are or aspire to be. This psychological interweaving is further supported by mirror neurons, which foster a sense of shared experience and empathy toward entities that reflect our values.

Aligning with a brand that embodies specific values is akin to making a statement about oneself without uttering a word – a phenomenon seen in the brain's medial prefrontal cortex where self-relevant processing occurs. This explains why a consumer might gravitate toward a brand like Patagonia, not only for its quality apparel but its staunch commitment to environmental protection, echoing their personal values of conservation.

Capitalizing on this notion, marketing efforts hinge on delineating a brand's commitment to social good – a charity event here, an eco-friendly initiative there – all tuned to evoke a synchronous neural choir of personal virtue and public responsibility. This alignment subjects the consumer's brand association to an ethical litmus test, with high marks fostering deep-rooted brand allegiance.

Yet, these value propositions must step beyond mere market ploys. Authenticity reads loud and clear through the social acumen regions of the brain like the temporoparietal junction, scrutinizing the genuineness of a brand's purported values. A savvy marketer ensures that a brand does not simply adopt values as a costume, but rather, that it lives them – in practice, culture, and narrative.

As consumers, we often jest at the transparent efforts of inauthentic brand virtue-signaling – our cognitive dissonance detectors sounding off like off-key notes in a symphony.

But on the other hand, brands become more popular when their ideals are easily integrated and align with people's personal beliefs (Wheeler 2017). Imagine brands that aren't selling goods but wearing beautiful clothes of virtues. Values are the threads that are carefully sewn into this fabric, telling the consumer's brain that they are in the right buying "tribe" in terms of moral and social association.

In the end, branding is a set of values that people think a brand stands for. Consumer-brand unity is both a mental and an emotional process. So, brand planners must walk a fine line between what the company offers and what customers expect, putting together a group whose members share values and a common goal.

Use GenAi to map out the overlap between brand values and consumer values, identifying key areas where the brand can authentically connect with consumers on a values-based level.

Deploy GenAi tools to analyze social media, reviews, forums, and consumer feedback for insights into the values, interests, and concerns of the target audience. This deep understanding allows brands to tailor their messaging to resonate with consumer values.

What Does It Say about Me to Others? Social Identity as Brand Benefit

In addition to the things they sell, brands are made up of a lot of different meanings and hints that tell us something about who we are. Brands speak eloquently on behalf of the consumer. The benefit of a Rolex is not accurate time – but rather to announce to the rest of the world, "I have arrived successfully." When we choose a brand from the many that are out there, it sends subtle messages to those around us about who we are and what our status is. In evolutionary biology, the idea of "signaling" means that actions or traits can send messages. In human societies, brands are one way that information gets sent.

A transaction is not simply an economic exchange; it's a transmission of social signals. This phenomenon, coined *conspicuous consumption,* turns the act of buying into a communication channel, where the products we own display our financial success, sophistication, and values. A person's choice of a fine Swiss watch, therefore, indicates more than just their appreciation for horology – it signals to others that they have reached a certain echelon in societal hierarchy.

Neuroscientifically, the benefits implied by brand possession activate the brain's reward pathways, involving the release of neurotransmitters such as dopamine which are associated with desire and satisfaction. This neurochemical cascade provides feelings of pleasure not only from the use of the product but from the societal messages it carries – the psychological allure of wearing a Rolex or holding the latest iPhone.

These brand-related benefits extend to the theory of self-concept and identity. The medial prefrontal cortex, a region implicated in self-referential processing, lights up when individuals evaluate brands that resonate with their perceived self-image. The brand becomes not just an object of desire, but a foundational piece in the mosaic of their identity – woven into the very fabric of who they perceive themselves to be, and who they wish to project to society.

What then does this mean for marketing? The straightforward implication is that marketing strategies should focus not solely on the features of a product but on the social and psychological benefits it confers. These benefits must be thoughtfully crafted to align with the desired self-image of the target audience, offering them a means to nonverbally broadcast their own personal narrative.

An example of this is evident in advertising campaigns that leverage the innate desire for social validation. Consider a commercial that whimsically portrays the envious glances a customer receives while using the product. By humorously highlighting this phenomenon, marketers tap into the social aspiration inherent in the consumer mindset – engaging them not only through the attributes of the product but through the emotive responses it's poised to conjure (Mathur 2016).

Aligning brand benefits with societal ideals necessitates cultural competency. A brand that signals ethical responsibility, for instance, takes on a halo of societal values that now resonate strongly with a consumer base increasingly conscious of corporate social responsibility. The subconscious message sent by using such a brand whisper of the consumer's commitment to these greater causes, striking a chord with their inner virtues.

The application of this dimension to brand communication requires a narrative that strikes balance between the explicit and implicit, ensuring that the benefits go beyond the overt sales pitch. When a customer selects a brand of electric vehicles, he's not simply choosing an energy-efficient mode of transportation – he's aspiring to join the chorus for environmental stewardship, a note of who he is and who he aspires to become.

However, embellishment should be approached with caution. Overzealous claims and benefits can lead to cognitive dissonance when the experience doesn't align with the promise. Hence, the benefits promised must be judiciously chosen and based on authentic and deliverable brand truths.

Understanding the layer of brand benefits is not only a way to get things done, but also a way to organize who you are. When marketers have a deeper knowledge of this idea, they can use it to weave benefits into an emotional tapestry where fun, ambition, and social norms all blend. The mixture must be both artistic and honest, because real rapport is what makes the greatest bonds between a brand and a customer.

Use GenAi-based consumer understanding to reveal the hopes, dreams, insecurities, and aspirations of the consumer. Identify how the brand may naturally speak to, and stand for the hopes, dreams, and aspirations of the consumer. Identify how the brand will dispel the insecurities of the consumer. Create content and messaging based on this analysis.

What Does It Say about Me to Myself? Personal Identity as Brand Metaphor

Through our choice of brands, we send messages to ourselves. These intensely personal and deep conversations occur in the nonconscious of our minds where metaphors are the language of choice. Metaphors are like clothes for the mind, and brands are the clothes we choose to show what we really believe and want. This powerful aspect of how people think gives brands the ability to be both a mirror and a light, showing who we are and where we belong in society. George

Lakoff and Mark Johnson, two linguists, have written a lot about how metaphors shape how we understand the world.

They assert that our conceptual system is metaphorically structured and defined, thus influencing not only language but also our thoughts and actions.

Through the metaphorical lens, a sneaker is not merely a shoe but might represent an extension of our athletic identity, an emblem of our dedication to health, or our penchant for casual style. This rich tapestry where products are imbued with meaning emerges from the way our prefrontal cortex engages in complex symbolic reasoning. This cerebral crucible melds brand symbolism with our self-identity and projects these values onto the societal stage.

To decode the narrative power of brands as metaphors, we can look to the default mode network (DMN), a constellation of interacting brain regions known for its role in self-referential thought (Soch et al. 2017). Here, brands ingrain themselves as metaphoric threads, woven into the fabric of the DMN, which in turn frames our sense of self and embellishes our unfolding personal saga.

Evoking affection or disdain, brands play upon the dopaminergic reward system, which fires up for more than just hedonic pleasure – it also lights up for meaningful stimuli, rich with personal relevance. When we encounter a brand that symbolizes sustainability, for instance, our neural circuitry buzzes with a sense of purposeful alignment if environmental stewardship is a fundamental value.

Translating these discoveries into marketing strategies, we delve into the construction of brand narratives that dress our aspirations. A brand is no longer a stationary construct but becomes fluid, dynamic – aligning with the metaphoric essence of our evolving identity, like changing outfits for different scenes in our life's play.

A car is more than a means of transportation; it's a vehicle that not only conveys you from point A to B but tells a tale of your environmental consciousness or your need for speed and luxury. The marketer's task is akin to that of a skilled tailor, fashioning a collection that allows consumers to adorn themselves with the desired metaphorical garments that speak without words.

When applying metaphor in branding, the goal is to align the brand's narrative so closely with personal aspirations that the consumer and the brand become virtually synonymous in the theater of the personal world.

However, as is the case with all powerful tools, the use of metaphor must be approached with the subtlety of a perfumer – a dash too much and the scent is overwhelming. Brands that lace their narratives with too heavy a dose of metaphoric grandiosity may find themselves the butt of consumer satire for overplaying their hand.

Use GenAi algorithms to generate creative and unique metaphors that align with the brand's identity, values, and the benefits of its products or services. GenAi can analyze vast amounts of data, including cultural references, consumer interests, and emotional triggers, to suggest metaphors that will resonate with the target audience.

Employ GenAi to ensure that the chosen metaphors are contextually relevant to the target demographic's culture, experiences, and values. This relevance increases the likelihood of the metaphor being understood and appreciated, enhancing its impact.

To Boldly Go Where No One Has Gone Before – Brand Extensions

Brand extensions echo the chameleon's talent, adapting and absorbing the essence of another to enhance its survival – in the brand's case, in the marketplace jungle. In psychology, the principles of assimilation inform us that individuals integrate new information by relating it to concepts they already understand. Similarly, when a brand extends itself, it leverages cognitive assimilation processes to blur lines and share the halo of an established brand's qualities.

Brand extensions are ventures into territories charted by others. They are acts of shaping perceptions using the priming effect, where exposure to one stimulus influences the response to another. By aligning with the characteristics of an admired brand, companies hope the consumer mind will be primed to receive their brand with the same adoration.

The neural correlates of brand extension can be understood through the lens of mirror neuron systems (Steward 2017). These neurons, which fire both when animals act and when they observe the same action performed by another, illuminate our capacity for imitation. In brand extension, the imitating brand hopes to echo the neural favoritism consumers have for another brand.

There is, however, a cerebral interplay here; a brand must balance novelty with familiarity. Extend too far and you might trip into the territory of incongruity, sparking a mismatch in the brain's expectancy, leading to cognitive dissonance. Ideally, an extension should be novel enough to stimulate interest but familiar enough to be believable and accepted.

From the marketing standpoint, the craft lies in strategically borrowing equity. The term *halo effect*, in social psychology, refers to the influence of an overall positive impression on the evaluations of a person's specific traits. A brand extension seeks to wear this halo by casting a beneficial light on its new ventures, enveloping itself in the attributes of the respected brand.

The extension process should not devolve into a clumsy masquerade. Witnessing a car brand suddenly leap into the world of confectionery, one might raise an eyebrow and chuckle; extensions must be credible and crafted with consumer psychology in mind to avoid being the clown at the masquerade ball. It takes finesse and deep understanding of branding narratives to move successfully into new domains.

This performance requires not only a deft understanding of communal neural tapestries but an intuition for public desire. The mind's association areas, located in the parietal lobes, are where sensory information converges into cohesive thought. A truly graceful brand extension taps into these convergences, dressing the product in a new costume while retaining the essential brand rhythm.

However, how these extended roles are communicated can determine the encore or exit stage left. The conceptual blending theory posits that new meaning is created by blending elements from different conceptual spaces. In branding, this means articulating the transition into new territory must be done in a manner that blends seamlessly with the consumer's mental models of both the old and the new brand identities.

Cognitive flexibility, the mental ability to switch between thinking about two different concepts and to think about multiple concepts simultaneously, is required of both the brand and consumer. Companies must navigate these cognitive acrobatics with a narrative that unfolds with humor and engaging stories that open minds to new possibilities (Braem & Egner 2018).

To conclude, brand extensions are a bid for consumer hearts that rely on cognitive science. Executed well, they can be an exhilarating display of agility and trust, which keeps consumers applauding for more. Within the ever-evolving realm of market spaces, brands that are capable of effectively assuming, adjusting to, and exuding fresh identities have the potential to garner enthusiastic applause.

Use GenAi to analyze market data, trends, and consumer behaviors to identify potential areas for brand extension that align with the brand's core values and competencies. This data-driven approach helps in spotting opportunities that have a higher likelihood of success.

Ensure that the messaging for the brand extension aligns with the overarching brand identity and values. GenAi can help maintain this consistency by analyzing existing brand materials and generating new content that seamlessly integrates the extension into the brand family.

Identify the Brand Archetype and Anchor the Brand

Brands deeply resonate with fundamental aspects of the human experience. A brand archetype influence serves as symbolic representations of a brand's persona that resonates with the latent narratives that are universally shared by all. The field of archetypes is intricately intertwined with evolutionary psychology and the collective unconscious, a concept described by Carl Jung as "collective unconscious" and representing the universal structural components of the human psyche. Like sacred symbols, brands elicit these archetypal images and derive their allure from the very neurology that connects us to legend and myth.

Every archetype, from the "caregiver" to the "explorer," narrates a unique story that connects with individuals on a primordial level. Neurologically, these connections may stem from our mirror neuron systems, which embody our capacity to understand and empathize with others (Lamm & Majdandžić 2015). These neurons could explain why certain brand stories feel so familiar and compelling, as if the brand were mirroring a part of our own internal tale.

The potency of archetypes is amplified by their ability to distill complex brand narratives into accessible and universal themes. Cognitive psychology recognizes the role of schemas – frameworks that organize and interpret information – asserting that archetypal themes simplify brand understanding through these preexisting mental structures. Consequently, a brand that effectively channels an archetype can enjoy the ease of aligning with consumers' preloaded cognitive maps.

The prefrontal cortex stands as an impresario of higher thought, orchestrating our ideas and actions with our social and psychological selves. This region hums with activity when we encounter a brand that genuinely embodies an archetype resonating with our internal values and aspirations, a harmonious resonance that strikes a chord with our personal and social identity.

As the cerebral narrative unfolds into practical strategy, marketers tap into the pulse of archetypes to compose a brand persona that feels both genuine and inevitable. The marketing maestro must gauge the emotional timber of the audience, selecting and refining the archetypal narrative that best resonates with the collective consciousness of the intended demographic.

The dynamic interplay between a brand's primary archetype and its product's secondary one creates a layering effect, much like a chorus accentuated by a solo performance that enhances the overall experience – each supports and enriches the other. However, it's essential that this duo perform a duet, not a duel; a campaign that awkwardly clashes with the archetype's essence may evoke a dissonant laugh rather than a standing ovation.

Brand spokespersons or social influencers selected to embody the brand's archetypal spirit are important choices. Choosing a face that captures the essence of the archetype is like casting the perfect actor for a role; their performance can enthrall or repel an audience, making or breaking the illusion of the character.

Use GenAi to analyze brand identity, values, consumer perceptions, and market positioning to identify the most suitable archetype(s) for the brand. This could involve mapping the brand's characteristics against the 12 classic archetypes (e.g., The Hero, The Caregiver, The Explorer) to find the best fit.

Leverage GenAi to analyze consumer data, including interests, preferences, and values, to ensure the chosen archetype aligns with the target audience's expectations and desires, enhancing the potential for emotional resonance. Employ GenAi tools to create compelling stories and messaging that embody the brand's archetype, weaving these narratives into advertising campaigns. This storytelling approach can make the brand's communications more memorable and emotionally impactful.

Demography and Age of the Brand – Brand as Person

Brands navigate the shifting time frames of public image as time passes. Brands change like living things, and their perceived age affects how their audience views them. Brand age, like the human life cycle, is important and temporal. Cognitive associations shape brand narratives over time, affecting relevance, dependability, and value.

A "grandfather" brand that has stood the test of time may carry with it a heritage of trust and dependability. This is etched into the fabric of consumer memory, with the hippocampus storing and recalling the longevity and consistency of a brand's narrative. However, a delicate balance must be maintained so as not to tip into obsolescence, remaining evergreen in the consumer's garden of choices.

Neuroplasticity, the brain's capacity to adapt to changes, parallels the need for brands to rejuvenate and innovate. Brands must adapt not only to the advancing clock but also to the shifting cognitions of their consumers, finding neural niches in the zeitgeist of each era (O'Mara 2017). This is comparable to synaptic pruning, where weaker neural connections are shed to make way for stronger, more relevant ones.

Yet embracing a younger, more "hip" persona can sometimes jar with established consumer expectancies. Dopaminergic activity drives pursuit and reward, and when a brand successfully navigates a youthful rebranding, it can tap into these reward mechanisms in the brain, inducing pleasure and desire within younger demographics. This pursuit of neural novelty can give a brand a fountain of youth, reinventing its cognitive connections with consumers.

Scientifically, the perception of age in brands activates preconceived social and cultural stereotypes embedded in our neural networks. These cognitive stereotypes can be as endearing as a genial elder or as dismissive as an "out-of-touch" antique. Marketers must ensure that their brand personas age like fine wine, growing more distinguished without sliding into irrelevance.

The voice of a brand must resonate with the tempo of its time while maintaining a coherent identity. Is it the time-tested wisdom that speaks to a loyal, mature audience? Or is it the fresh, vibrant energy targeted at the restless heartbeats of youth? Put another way, in the big brand ballpark, is your team wearing vintage jerseys or innovative athletic gear?

But tread lightly, for a brand taking itself too seriously in the process of rebranding may have an identity crisis. Authenticity, always a prized commodity, can become lost in translation when a brand attempts to leap generations without a narrative safety net.

However, brands can gracefully sashay down memory lane, using the past as a launchpad rather than a crutch. Consider the renaissance of vinyl records or classic fashion making a comeback – trends that engage the nostalgia network in the brain, a delicate dance between the substantia nigra and the amygdala, eliciting warm, fuzzy feelings alongside the allure of antiquity. As brands progress through time, they must retain their essence while matching the market's ever-changing rhythm – one that respects tradition while tapping into the beat of modernity.

The goal is to cultivate a brand persona that is comfortable in its chronological skin, adapting to the consumer's ever-evolving mental landscape. As cultures change and consumers evolve, those brands that evolve and change with the consumers live to dance another day.

GenAi can identify the demography of the brand – the age and cultural identity. This information can then be intelligently used to create brand content that is age and culture appropriate.

Crafting Brand Sonic Signatures – Voice of the Brand

The auditory landscape of our lives is intricate – the rustling leaves, the thrumming cityscape, and among them, the resonant frequencies of our favorite brands. Sonic identity refers to the auditory components that articulate a brand's essence through melody, rhythm, and texture. Just as birds are identified by their calls, a brand's sonic signature is its call to its audience's memory and emotions. Neuroscience opens the stage for us to understand how our brains process and attach meaning to sounds, revealing the insulin-like growth factor 1 as one of the hormones involved in auditory learning mechanisms (Murillo-Cuesta et al. 2011).

Auditory stimuli interact with the auditory cortex, eliciting not only recognition but complex emotional and associative processes. A well-crafted sonic identity transcends mere jingles, embedding itself within the neural networks responsible for long-term memories – the entorhinal cortex playing a key role in this musical map-making process.

To examine the archetypal impact on a brand's sonic identity, we see Jungian psychology interfacing with auditory processing – a hero's fanfare, a caregiver's lullaby. Each archetypal melody threads itself through the fusiform gyrus, an area significantly involved in complex cognitive processes like perception and recognition. The choice in timbre, pitch, and rhythm speaks not only to the brand's character but echoes through our cerebral corridors as a known mythic figure.

Auditory branding starts its own journey through time, just like a face either ages gracefully or withers away from lack of care (Minsky & Fahey 2017). Sonic identity needs to keep its tonal quality fresh in the hippocampal rooms of group memory to stay relevant. How a brand adjusts its old tune to the changing tastes of people requires careful renewal and steady consistency.

When it comes to practice, marketers recognize the sonic branding as a touchpoint – an audible key to the consumer's associative palace. Just as the olfactory bulb skirts the rational brain to trigger emotion, so does sound. A marketer's task becomes akin to that of a composer, understanding the brand's identity to create an audio piece that embodies its very essence – ensuring the consumer not only hears the brand but feels it.

Brands with distinctive audio marks find themselves nestled within the amygdala's embrace, associated with affect and memory.

Imagine if a timeless brand like Coca-Cola, instead of catchy pop music, opted for Gregorian chants. The public might applaud the bold move or just as likely recoil in bemused horror. The same applies to sonic branding – while the familiarity of heritage has its charm, relevance requires the brand to sway to contemporary beats.

The sonic realm for brands isn't merely about carving out a cognitive niche; it's about harmonizing with the consumer's life song. Each echo of a brand's melody joins the consumer's daily rhythm – whether it's the playful buzz in a gadget store or the soothing whispers in a spa. It's not about the loudest sound; it's about the right pitch at the right time, sung with an air of effortless authenticity.

Use GenAi to generate unique sonic signatures based on the brand's identity, values, and target audience preferences. GenAi can analyze vast datasets to produce sounds that are likely to resonate with the intended audience while ensuring the signature is distinctive and memorable. Employ GenAi algorithms to adapt the sonic signature for different cultural contexts, ensuring it maintains its appeal and relevance across global markets. This includes adjusting musical scales, instruments, and rhythms to suit local musical preferences.

Leverage GenAi to create interactive advertising experiences where consumers can engage with the sonic signature in creative ways, such as remixing the brand's jingle or incorporating it into user-generated content. Interaction increases memorability and emotional attachment.

In this chapter, you gained knowledge and AI tools to build effective brands – brands that last, brands that consumers love, and adopt as their own.

You added "sonic signatures" to your roster of brand wisdom. And lessons about why metaphors matter so much. The nonconscious mind's secrets to effective brand extensions were shared.

Finally, you were given direction on how to enlist GenAi in your quest for brand supremacy. This will become the modern defining difference in determining which brands grow and prosper, and which brands stay stagnant.

CHAPTER 12

trendGPT – Discover Trends and Product Innovations Using neuroAi

D espite a cottage industry of experts who claim expertise in forecasting trends, the ability to see over the horizon remains an elusive, even mysterious goal – until now.

We all had to have a bit of faith and grab a bucket of popcorn to watch the movie of the future.

The advent of AI has brought unprecedented power to the art of trend identification and by extension, product innovation. It has made it a science – and one that you can master. Combined with the vast storehouse of knowledge that is modern neuroscience, AI is offering the capability to identify emerging trends before the competition. That extends out to specific geographies and demographics.

Gaining a grasp of this innovation phenomenon will be the key to business success going forward.

Begin with a Big Bang: Start with the Consumer – Profile, Needs, Irritants

The first step toward innovation is to begin with the consumer. Translate a simple definition of the consumer into a detailed sketch based on age, gender, country, and culture.

Delve into nonconscious data such as the elements found in songs, movies, and TV shows to create a compelling portrait of the consumer.

GenAi algorithms can be fed simple consumer descriptions from which the salient descriptive elements of the consumer are derived from the nonconscious corpus related to the consumer. A detailed description of the archetypes, hopes, dreams, and desires of the consumer are derived and extrapolated.

GenAi algorithms can effectively build a profile of the consumer. The way to train algorithms is to get them to reflect on the sparse information provided, and ask for implications of what is provided. Algorithms must be asked to connect pieces of information together and draw implications. These implications can be further connected to draw secondary and tertiary implications. What does living in a city, and being single mean? Algorithms must be taught how to interpret facts and data to draw conclusion about what needs, frustrations, hopes, dreams, desires, disappointments might be for a consumer in their professional, personal, and social life. n-grams of interpretations build richer consumer profiles. Then algorithms must comb for consistency and flag outliers. It is important for algorithms to identify the little points of "friction" in the consumer's life. These are not major needs but the minor irritants that get in the way of enjoyment or full expression. Understanding these minor irritants, and fixing them, can lead to major innovations and market success.

We posit that true innovations rise from these descriptions of the consumer and a deep understanding of what their needs are. Traditionally, market research led the way here, and today algorithmic and programmatic research supplements and at times supplants traditional approaches.

A calculation is performed to determine the price sensitivity of the consumer to the product and category in question. This will serve to perform volumetric analysis later. Innovations must bear in mind the willingness of the consumer to pay. So algorithms must have a general idea of whether the goal is to create a product, premiumize it, or create a new category. Without the price sensitivity, innovations will never facilitate consumer innovation.

Emotional Memories: Trends and Innovations That Rise from Feelings

Emotions serve to color our memories on the mental canvas known as emotional memory. It is the central point at which feelings evoked by events converge with experiential details to form a multifaceted, nuanced memory. Emotional memory is the vivid hue that might persist long after the event has faded, in contrast to the dry data points of semantic recall. It's what makes us laugh years after a humiliating mistake or savor the nostalgia of an era long gone.

At the neurobiological core, the amygdala plays the starring role in encoding and retrieving emotional memories. This small, almond-shaped structure helps to modulate memory consolidation, acting as the mind's emotional thermostat, turning the heat up or down on our recollections based on their emotional intensity. Heightened emotions signal the brain to plant a memory firmly in this fertile emotional soil.

The unique aspect of emotional memory is its ability to bypass the typical routes of conscious recollection. We may not remember the mundane details of a day that brought us joy or sorrow, but the emotional tone of that day is etched into our very being. In marketing, tapping into these effectively charged networks can mean the difference between a product that is easily forgotten and one that lives on in the consumer's heart.

The concept of *emotional branding* capitalizes on this natural encoding of emotional memory (Bhatia 2019). Brands that story-tell with emotional arcs resonate on a more personal level, knitting themselves into the fabric of life's highs and lows. It's like that one-hit-wonder song that instantly transports you back to your teenage years, complete with all the angst and butterflies of first love.

When connection to a product or service transcends its utility and weaves itself into the personal narrative of the consumer, it achieves a form of emotional immortality. Consider the furor of fans defending their beloved smartphones or sports shoes; it's not just about gadgetry or apparel – it's an emotional allegiance.

This is not to say that brands should manipulate emotions carelessly. Authenticity is the guardian of emotional memory. Transparent and genuine brand stories foster trust, whereas disingenuous attempts at emotional manipulation can leave indelible stains of mistrust on the fabric of memory.

Rituals and traditions are powerful activators of emotional memory (van Mulukom 2017). Think of the holiday season commercials; they rarely speak of the product but of the feelings that surround the holidays. They bank on the collective emotional memory – that comforting nostalgia that rolls around each December, urging us to relive and recreate those special moments.

As we consider how emotional memory paints the experience of innovation, it's clear that products that evoke feelings, that tug at the heart, are more likely to be cherished, shared, and remembered. Whether it is soothing the irritation of a poorly designed interface or the exuberance of discovering a product that feels tailored to one's exact needs, the emotional responses become part of the product's essence.

Employ GenAi to predict emotional trends within your target demographic or industry. By staying ahead of emotional shifts, you can tailor your advertising messages to align with the evolving emotional landscape of your audience.

Sensory Memory: Trends That Rise from Sensory Expectations and Disappointments

Sensory memory is a fleeting reservoir of sensory information, a kind of ultra-short-term memory that preserves the richness of our perceptual experiences for mere seconds. It's where the shimmer of a visual, the faintest whiff of a fragrance, or the vestige of a melody first make landfall in the human mind. The breakdown is granular, with distinct memory stores for each sense: iconic (visual), echoic (auditory), and haptic (touch), among others.

The intricate pathways of our senses feed into this ephemeral storage, ruled by the principle of immediate perception. Sensory memory acts as a buffer, holding incoming stimuli just long enough for processing and potential transferal to more enduring forms of memory. It's our neural snapshot, crisp and vivid, giving us time to decide what's worth saving in the more permanent album of the brain. Sensory memory is both memory and activation – a view of the world and navigating it through powerful, ancient, and rudimentary mechanisms of smell, taste, touch, imagery, and audition. Marketers intuitively understand the power of scent, texture, fizz, and jingle.

Visually, iconic memory lets us perceive a seamless stream of imagery despite the rapid eye movements that could render life a chaotic flicker. Quick glimpses are enough for the brain to capture and briefly retain detailed information, turning a fraction of a second into a lingering gaze in the mind's eye to be sifted and sorted.

Echoic memory, on the other hand, helps us retain auditory information after the original sound has ceased. Our brain keeps echoing the last few words of a sentence long enough to extract meaning and establish context – like the cerebral equivalent of an audio loop. It proves essential in understanding speech, where syllables tumble quickly and order matters.

Our senses of smell and taste come with their ephemeral repositories. These gustatory and olfactory impressions are strong emotional triggers, tied closely to the visceral and emotional brain areas, such as the amygdala. A scent can transport us back to childhood kitchens; a flavor can rekindle forgotten summers.

Sensory memory is fast paced. Yet, the precision of those initial seconds – when a consumer first encounters a brand – can determine whether an experience becomes ingrained. Marketers are the choreographers of these fleeting moments, creating sensory experiences that strive to harmonize with emotions and become lasting memories.

Sonically, the jingles and sound logos stamp an audio fingerprint in the echoic memory – a tune hummed long after the commercial ends. It's a whisper of brand identity, as lighthearted as a catchphrase from a favorite sitcom, that repeats in the mind's echoing chamber, striving to become a part of the viewer's internal playlist.

And who could ignore the olfactory logos, the signature scents wafting through boutique doorways, or the zest of a citrus-infused product? These aromatic trademarks are less about the nose and more about the emotional snapshot they seal within our sensory archives (Gvili et al. 2018). GenAi algorithms utilize sensory memories to implant features and attributes to products and innovation. The satisfying thunk of the door when it closes, the click of a bottle cap that lets us know that a product is safely stored, the little extra texture in the grip of a handgun, the fragrance in a bag of chips when it is opened, the feel of bubbles in a cola at the first sip – all are innovations that excite the sensory memories in the consumer.

A World Constructed: Trends That Rise from Semantic Memories

Semantic memory is the archive of our accumulated knowledge, distinct from the personal episodes of our lives. It comprises the cold hard facts, concepts, words, and relationships among them that we draw on to understand and interact with the world. This dimension of memory is like a vast encyclopedia within our brains, filled with entries sorted not by personal experience but by meaning and connection.

Neuroimaging studies reveal that semantic memory is widely distributed across the cortex but has focal points in the temporal and parietal lobes. This intricate web, with neural nodes and networks, is where language and cognition intertwine. It empowers us with linguistic constructs to label the world and imbue it with a shared meaning. Semantic memory on the other hand is related to views and perspectives of the world, skills, structure and organization of knowledge and language.

Semantic priming exemplifies the interconnectivity within this system. When we hear the word *nurse,* our brain simultaneously activates related concepts such as *doctor, hospital,* and *care.* This underlying network paves the way for associative learning where one idea effortlessly leads to another, a feature that clever marketers can deftly exploit.

Conceptual knowledge in semantic memory develops through rich encodings of multisensory information which become abstracted over time. Such abstractions become the building blocks for complex thought. They transcend specific experiences, allowing for the generalization and application of knowledge across different contexts – and therein lies a marketer's playground.

The theory of spreading activation suggests that thoughts weave through this web of memories, lighting up pathways as they go. For marketers, understanding this process is akin to becoming a puppeteer who can subtly pull the strings to guide the audience's thought patterns.

Brand narratives can become cultural tenants within an individual's semantic network (Allen et al. 2018). Slogans such as "Just Do It" encapsulate more than a brand directive; they tap into universal constructs of motivation and resilience that resonate broadly across semantic memory landscapes.

Semantic memory is in constant flux, accruing new information and forging new connections. A dynamic brand identity, one that evolves through campaigns while firmly rooted in core values aligns well with this ever-shifting semantic ground. Such adaptability reflects the plasticity of the cerebral code itself.

Employ GenAi algorithms to map semantic triggers to specific contexts or consumer needs, ensuring that the chosen features and product attributes are attuned to preexisting semantic memories.

Memorable Marketing: Trends That Harness the Power of Episodic Memory

In the human mind, episodic memories are the captivating dramas, the moments imbued with emotion and personal context that shape our narratives. These memories are a record of our experiences, akin to vivid scenes in the episodic structure of a television series, forming what scientists call our episodic memory. Every interaction, every touchpoint a brand has with its consumers, crafts a potential episode in the grand series of their lives. Episodic memory connects to experienced events, occasions, and life experiences.

Neuroscience elucidates how episodic memory is more than a mere mental replay; it's a reconstruction, a dynamic and malleable storytelling process within our brain (Dings & Newen 2023). Each recall is an opportunity for embellishment, a chance to infuse the narrative with new details that brands can influence. It's an insight critical in understanding how product innovation can go beyond transactions to become part of the consumer's life story.

The creation of episodic memories is subject to a complex neurobiological process. The hippocampus directs the formation and retrieval of these stories. Its role is pivotal, and any marketer's aim should be to stage experiences that warrant the hippocampus's attention, making these moments conducive to creating lasting episodic memories.

Note that temporal memory is an integral component of episodic memory. The time sequence and order of events and things as they occurred provide crucial contextual information. This temporal memory serves to illustrate various points of joy and friction in an experience. Indeed an experience can truly be summarized by its emotional high and low points and the order in which they occurred.

There is an inherent emotional component to episodic memory. Emotions often dictate the intensity and durability of these memories (Beard et al. 2024). Marketing that resonates emotionally can, therefore, expect better memory retention. This is where narratives that stir laughter, joy, sadness, or even surprise can transform an ad into an enduring memory and a point of profound connection with a brand.

Yet, creating episodic memories isn't about raw emotion alone; it's also about relevance and relatability. Consumers are more likely to remember and cherish experiences that align with their personal narratives and values.

But how do we ensure that the episodic narrative endures beyond a campaign's crescendo? Feature sets and product innovations, but of the subtle kind, reinforce these mental storylines. Product features that clearly eliminate friction, and consumer irritations that's woven through various forms of feature sets, interactions, and experiences ensure the innovation uniqueness won't fade when the curtains close on a particular marketing campaign.

As episodic memories can be revised with each recall, there's room for brands and product innovations to become integrated within these episodes securely. Through strategic innovations and feature sets that inspire self-generated thought, a product becomes more than a prop – it's a character with a recurring role in the consumer's episodic memory, adding depth and connection to the brand-customer relationship.

Consumer experiences evoke these memories, serving as anchor points throughout the marketing funnel. The points become episodic markers, staging the potential for new memories. By harnessing significant events, marketers can craft product experiences that become natural extensions of the consumer's personal narrative.

Use GenAi to generate innovations that are designed to evoke strong emotional responses to consumer hopes and pain points. Incorporating trends and innovations into these product interactions in a way that feels personal and relatable can help anchor the brand within consumers' episodic memories.

Utilize GenAi to incorporate specific sensory cues into products that can trigger episodic memories related to trends, such as the smell of rain associated with eco-friendly products, or the texture of a material linked to a new fashion trend.

The Geocoding of Experience: Trends Connected to Spatial Memory

Spatial memory is woven into the fabric of our minds, representing environments from the grandiose to the quotidian. This aspect allows us to navigate our world, remember routes, and find our car in a parking lot. The hippocampus, our neural GPS, specializes in storing and processing this information about location and spatial environment. It captures the layout of rooms, buildings, and broader landscapes, encoding them within an internalized map of our experiences.

Beyond its navigational utility, spatial memory hosts a wealth of associations. From the classroom where pivotal learning first clicked, to the cozy cabin in the woods of our fondest getaways, these "mindscapes" are not just coordinates but emotional beacons. They mark the geographies of our hearts and minds, each associated with its collage of feelings, smells, and sounds.

Our encounters with places engage a multitude of neurological actors. The place cells in the hippocampus fire in specific locations, encoding the "where" of our stories. Meanwhile, the entorhinal cortex's grid cells provide the brain with an internal metric to gauge distances and directional bearings. Together, they create a multidimensional spatial mnemonic that serves as the backdrop to our personal narratives (Moser et al. 2015).

Turning our gaze from the biological landscape to the terrains of marketing, brands often leverage spatial memory to anchor their identity. Whether it's the anticipation of ambiance just by seeing the golden arches of McDonald's or the distinct Scandinavian aesthetic of IKEA showrooms, businesses tap into the spatial dimension to trigger a whole suite of associations and emotions.

Indeed, spatial triggers in product innovation go beyond mere location – they are the lighthouses guiding us to the reefs of consumption. A familiar jingle may transport us to the aisles of a supermarket, or the scent wafting from a bakery can pull us back into the warm Sunday mornings of our childhood kitchens (Chu & Downes 2000). It's this navigational interplay between sensorial cues and spatial memories that brands can choreograph for magnetic resonance with consumers.

Yet the spatial realm is not only external but also highly internal and imaginative. In branding, one can employ the concept of a "mental showroom," where consumers' spatial memories are embellished with aspirational experiences tied to products or services. The journey within a premium car commercial doesn't just exhibit an automobile; it takes the viewer through winding roads and luxurious cityscapes, animating the viewer's spatial memory and embedding the brand within it.

Retail innovators often chart out consumer's spatial narratives within retail spaces. The deliberate design of store layouts influences not just the path taken, but the memory of the experience. It is not uncommon for shoppers to possess a spatial narrative for their favorite store, a kind of choreography directly affected by the spatial structuring within (Batey 2015).

While the hardwiring for spatial sensing is universal, the personal gloss on these places is unique, which presents an untapped expanse for personalized marketing. Imagine a company creating virtual reality experiences that simulate visiting a bespoke boutique in Paris – these tailored "place-memories" could be the new frontier in digital marketing.

Spatial memories become important in product design – products and features that evoke journeys through mental landscapes of places, sights, and scenes – either known or imagined.

Ritual and Habit: Procedural Memory in Product Innovation

Our mechanical intelligence is stored in procedural memory, which is also the quiet orchestrator of tasks like riding a bike, playing an instrument, and even typing without looking at the keyboard. Procedural memory, in contrast to episodic or semantic memories, is the conscious recall linked to our automated abilities and routine behaviors. It functions as a sort of cognitive autopilot, functioning below the level of conscious awareness.

These processes are largely managed by our basal ganglia and cerebellum, playing a critical role in the development and execution of motor skills and habits (Graybiel & Grafton 2015). This part of our neural machinery allows for physically and psychologically complex tasks to become second nature through repetition and practice.

It's not just motor skills; cognitive routines – be they mental checklists before leaving the house, or the procedural journey through websites we regularly visit – also fall into the remit of procedural memory. These sequences, once established, may require nominal to zero conscious effort, freeing up our cognitive resources for other tasks.

The principles of conditioning are deeply entwined with procedural memory. Pavlov's classical conditioning experiments come to mind, where a neutral stimulus, through association, comes to elicit a behavioral response (Pavlov 1927). Through repetition, a brand jingle or logo can become as automatic in triggering consumer recognition as the smell of coffee is in signaling the start of a day.

In terms of habits, context cues play a part in shaping behavior. Context-dependent memory from the textbook of marketing psychology tells us that the environment in which we learn information can trigger the recall of learned behaviors. Brands that are skillfully enmeshed in daily routines become as unthinking a choice as grabbing the familiar toothpaste in the morning rush.

Behavioral therapy has shown that new habits can be grafted onto old ones through repetition and positive reinforcement. Applying this to branding means creating positive experiences around products or services that can pave the way to habitual consumption. Reward, in this sense, comes not just with usage, but with the consistency of the message and its repetitions.

What happens when a routine is disrupted? The brain is a lover of patterns but also has plasticity, providing the flexibility to adapt to new behaviors and procedures. A jarring break in a commercial, or an unexpected twist on a label can reset the procedural narrative, making room for new habit-forming rituals that align with a brand's marketing strategy.

Utilize GenAi to analyze consumer data and identify common daily rituals and routines among the target audience. This could include morning routines, mealtime habits, exercise regimens, or evening rituals. Products, and features must naturally fit into these routines, or be able to create newer rituals for consumers. Note that newer rituals become harder to adopt, but a product that fits into an existing ritual is more easily adopted by the consumer.

Memories and Metaphors: Predicting Trends before They Happen

Trends might feel like they appear out of the ether, taking hold of society in a sudden rush. Yet, the rise of global megatrends is anything but mystical. It concerns the predictable streams of changes that glide across borders and cultures, painting the future landscape of consumer behavior. As unique as these forces are, they seldom occur in isolation, and often are rooted deeply in the collective psyche, which neuroscience is beginning to unravel (Alsharif et al. 2023). Let's explore the epicenters of these megatrends, and how, at their core, they connect to our brain's wiring.

Seismic shifts in demographics, technology advancements, environmental concerns, they all serve as pillars holding the architecture of megatrends. The first task of a marketer is hence akin to an archaeologist – uncovering the impact of these vast movements both globally and within the minutiae of local cultures and sub-demographics. This exercise is not merely academic; it's the fertile soil in which innovative marketing strategies take root.

But understanding trends is just the beginning. The second step involves delving into *memory structures,* those neurological powerhouses where experience and brand interactions take up residence. Neuroscience tells us that our memories aren't just passive repositories but active, dynamic players in how we perceive and react to new information. In trend innovation, constructing memory palaces with connecting hallways to these structures can make the difference between a passing fad and a lasting movement.

The various memory structures outlined earlier – episodic, spatial, sensory, semantic, procedural, and emotional memories are store houses of the next generation of trends and innovations. Events in the world bring what is already deeply locked and rooted into focus and sharp relief. GenAi algorithms must neatly catalog the vast set of factors contained in memory structures, and must relook at them through the lens of culture, geography, and world events. This automatically enables us to spot trends algorithmically in a timely manner, and get a little ahead of everyone else.

The analytical mind engages, dominating the narrative mind, and consequently, emotional engagement sputters – a fantastical story reduced to bullet points. Functional magnetic resonance imaging (fMRI) studies support this, showing distinct brain activation patterns when processing stories as opposed to logical, feature-based information. Consequently, keeping the story alive, even in the backdrop, is crucial to maintain engagement.

But how does one keep the narrative flames stoked amidst the specifications and features of products? It's gentle nudging as opposed to pushing, weaving a story subtly as the fabric of facts unfurls. Visual cues act as storytellers, maintaining an emotional connection even as intellect takes the wheel. The presence of relatable human elements within marketing content signals the brain to produce oxytocin, the so-called love hormone, which fosters a feeling of trust and connection.

Linguistic metaphors serve to confirm algorithmically predicted trends – the nonconscious language of the brain, and its equations translated nicely into trends. Emergent metaphors become the trends in products and services of tomorrow. Our language contains within itself the seeds of the desires of tomorrow. Memories and metaphors generate the trends that are just around the corner.

The Cross-Pollination Alchemy: Innovations Leap across Categories

Innovation and cognition spin on the same axis, revolving around an intricate network of memory structures that are interwoven with daily experiences, habits, and sensory triggers. The science of memory is not a monolith; it's a multifaceted crystal, reflecting different cognitive processes depending on the light shone upon it. From the recall of episodic events to the inveterate patterns of procedural routines, our brains maintain a cornucopia of information that brands tap into for innovation.

By marrying these memory dimensions with the conditions and triggers such as music, fragrance, and scenery, brands can stitch new sensory threads into the fabric of familiarity. Impressions in sensory memory, though fleeting, leave behind an echo that can reinforce or redefine emotional connections, an integral part of the trend-setting process.

Taking these threads together, we see innovation, where mega trends meld seamlessly with human experience and memory structures to create products that resonate. It eschews the traditional, embracing a palette of experiences painted with the vibrancy of city life and the calm of green spaces. By embedding innovation into this aesthetic, a brand captures the zeitgeist of modernity while remaining connected to the visceral and collective memory of its audience.

Within this terrain, neighboring classifications frequently meld together via the synesthetic orchestra of our sensory matrix. A scent could inspire an image; a tune could announce a taste sensation; these innovations push the limits of what a product can represent, like when technology and outdoor design come together to create devices that look just as good on a desk as they do in a backpack.

The artistry of innovation lies in leveraging and extending these memory rooted trends into new domains, reaching into distinct categories to bring back ideas that spawn the Aha! moments. The alchemy of analogy that takes wins and surges in one category and tastefully predicts its appearance in a related but distant category. Like embeddings trends and innovations cluster together in a higher dimensional space. It might be the reinterpretation of a retro fashion motif in modern apparel or the reimagining of a traditional ritual with state-of-the-art technology – all while keeping a firm grip on the emotional and cognitive coattails of the past.

Innovation arises naturally when trends in one category slowly diffuse into adjacent categories – the trends in soups make their way into teas as both are consumed when one is not feeling well. So cross-category, cross-pollination is critical for innovation and trendspotting. Human ingenuity is just this beautiful diffusion of knowledge, sights, and sounds across categories. Breakthroughs in science and design are as simple as taking insights and understanding in one area and applying them in a different domain or seeking their analogs in a different domain. GenAi algorithms can accomplish this beautifully, seamlessly, across a number of categories.

As GenAi algorithms create product innovations, they must follow a template that allows one to identify how the innovation was generated. The template can include the following elements: Name of the Product, Catchy Tagline that captures the essence of it, details of the innovation, the memory structures that are the source of the innovation, the consumer hope, pleasure, pain, or minor friction the innovation seeks to address, the context and circumstance of the innovation, context of the consumer memory structure, the functional connectivity of the innovation to the consumer's life, the criteria the consumer may use in judging the relevance of the innovation. Algorithms can automatically score and rank the innovation based on criteria such as functional benefits, emotional benefits, depth of connection to consumer memory structure, consumer expression of individuality, shareability and talkability, level of engagement of formats, sensory appeal, stickiness, and desire for regular use by the consumer. These scorecards can be customized for each company, each category and sub-category of product.

In this chapter, we have identified how product trends can be spotted and product innovations identified using a combination of neuroscience and GenAi. We have also recommended how cross-category pollination and analogies are critical methods to create innovations and spot trend spillovers. We see the next generation of IP and patent factories that generate not only innovations but patent applications that can protect them.

CHAPTER 13

packageGPT – neuroAi-Powered Package Design

P ackaging has always been more art than science.

That has just changed, due to the blending of advanced neuroscience with GenAi.

In this chapter a number of neurological insights will be disclosed, ranging from the importance of numbers on a package and how they must be controlled to the types of typography that make for the most effective packaging designs, the rule of threes, and much more.

How to put pop-outs to use, the importance of texture, and the critical role of processing fluency will be revealed.

Turn the pages for packaging proficiency for your business.

Packaging That Catches the Mind's Eye

The retail shelf communicates through a tacit language, comprehended on a subconscious level, through an interaction of visuals and text. The arrangement of these materials might be considered both an artistic endeavor and a precise scientific process, due to their intrinsic beauty. The interaction between pictures and words may be described as working in harmony.

This interplay begins in the brain, where the left and right hemispheres divide duties like office coworkers specializing in distinct roles. Neuroscientific research confirms this cerebral segregation: the right hemisphere excels in processing visual and spatial information, while the left hemisphere is superb at verbal and analytical tasks (Gerrits et al. 2020). This natural partitioning sets the stage for packaging design, creating a spatial framework where the placement of imagery and text is more than just aesthetics; it's a cognitive handshake with the mind.

Understanding how the brain works influences how we understand and engage with product packaging. The journey of imagery from the left visual field to the right hemisphere is smooth, a direct line to the domain of image interpretation. Conversely, text is more readily processed when presented in the right visual field, immediately accessible to the analytically inclined left hemisphere. Acknowledging this divide is step one in transforming packaging from mere containers to communicators.

The cognitive fluency factor shapes the ease with which we interpret and internalize information. When images rest on the left side of the product packaging and words on the right, cognitive fluency is maximized – our brains absorb the information efficiently, with minimal effort and maximum pleasure. This resonance of fluency taps into the cognitive preferences wired within us, enhancing the appeal and clarity of the product's message.

If the brand logo, the symbol of an identity, resides on the left, it benefits from immediate and unencumbered image processing (Spence 2016). Yet, when centered, it commands undivided attention, monopolizing the brain's focus through its prominent positioning. Each option, left or central, conveys a unique narrative while maintaining seamless cognitive engagement.

Translating neuroaesthetics to shelf presence magnifies a product's charisma in the competitive marketplace. Integrating neuroscience suggests a left-imagery, right-text strategy that plays to the strengths of our neural processing. The artful execution of this design does more than catch the eye – it respects the cognitive mechanics at play, transforming the packaging into a tapestry that speaks directly to the subconscious mind.

Humorous images or witty text touch upon the brain's reward centers, offering a moment of levity that breeds connection and recall. A lighthearted approach woven into the design invites a relatable human touch amidst the precise science of placement.

Utilize GenAi to generate symmetrical patterns, shapes, or motifs that can be incorporated into the packaging design. Symmetry can create a sense of balance and order, making the packaging more visually appealing to consumers.

Employ GenAi algorithms to ensure that the layout of design elements on the packaging is aligned with the neuro principles mentioned above. This promotes a cohesive and organized appearance that enhances consumer perception.

Category Busting Metric: Winning with a Single Number

The landscape of a package is often a mosaic of figures, from calories and prices to weights and dates. Science dives deep into the human brain's relationship with numbers, discovering a realm where cognitive load and number processing intersect. The prefrontal cortex takes the reins in numerical cognition, regulating our ability to understand and manage these mathematical symbols. The simplicity of single number presentation aligns with the brain's preference for minimal cognitive workload, fostering a smoother journey to comprehension.

Cognitive capacity and the numerical overload is an area abundantly researched. Cognitive psychologists propose that our mental bandwidth – the working memory – is finite and easily swamped by an excessive array of stimuli. The excess of numbers on a package can lead to cognitive overload, a state where the consumer's ability to process and retain information is compromised, making the path to decision-making cluttered and exhausting (Chee & Goh 2018).

The isolation effect, or Von Restorff effect, proposes that in a group of similar objects, the one that differs from the rest is more likely to be remembered (Von Restorff 1933). Applying this to numerical information on packaging, highlighting a significant figure allows it to stand distinct and memorable, rising from the sea of numerical data to anchor itself firmly in the consumer's memory.

Numerical alignment and pattern recognition play to the brain's love for order and predictability. Studies in perceptual organization suggest we are innately drawn to patterns and sequences, finding comfort and ease in the predictability they afford (Palmer & Rock 1990). A neatly arranged suite of numbers permits the brain to process numbers with greater ease and higher efficiency.

Neuroaesthetics of minimalism has its principles rooted in the movement toward reducing cognitive effort. The minimalist approach in design, which translates to the use of fewer figures, aligns with our neurological inclination toward streamlined and uncomplicated visual fields. This neural penchant for simplicity is why a solitary, salient number on a package can be the lynchpin for effective communication with the consumer.

Applying numeric focus in marketing begins by identifying the most pivotal piece of numerical information for the consumer and enshrining it on the packaging. Whether it's the caloric count for health-conscious buyers or the price point for value-oriented shoppers, plucking this number from the numerical nexus and granting it the spotlight simplifies the consumer's decision-making process.

Be it megapixels in a camera, gigahertz in a computer chip, or the number of lenses on a phone, the brain seeks a single number to differentiate a product. The package must honor and reveal that number in bold at the point of purchase.

When multiple numbers vie for attention, none emerge victorious. However, when one figure is left to stand tall, unchallenged in prominence, its impact is outsized – a singular testimonial to the importance of "one." This approach ties into the principle of salience, where the unique element in a context of similarity stands out and grabs the limelight (Itti 2007).

Use GenAi to design packaging that prominently features the single number claim, ensuring that it is immediately visible and eye-catching to consumers. This could involve placing the claim on the front of the package in a large font size or using visual cues to draw attention to it.

Incorporate GenAi to extract the category busting metric. Utilize the features of the product and the steps in the process of producing the product, and map it to a detailed analysis of the consumer to determine the key metrics for salience. Emphasizing how the product addresses a specific need or solves a problem can increase its appeal to consumers.

Typography Tango: Funky Fonts, Verticality, and Orientation

Steve Jobs studied calligraphy at Reed College. He understood the value of fonts when Apple made its first product. Typography steps onto the stage with poise and intention, embodying the silent yet powerful medium through which a message appeals to the consumer's consciousness. Neuroscience reveals that from the intricate network of neurons a fascination with typography emerges, rooted in the brain's response to visual stimuli. The fusiform gyrus – a region tasked with the processing of color, form, and faces – is also pivotal in recognizing typographic details (Weiner & Zilles 2016). This cerebral sensitivity to typography underscores the potency of fonts as vessels that not only carry meaning but also engage the brain's aesthetic sensibilities.

The visual word form area (VWFA), nestled within the left fusiform gyrus, becomes activated when we encounter printed words, hinting at the importance of font choice in communication. A typeface is not merely a garment for words – it influences legibility, evokes emotion, and contributes to brand identity.

High cognitive load springs from the challenge presented by overly ornate or complex fonts. These elaborate characters may push the boundaries of artistic expression but simultaneously demand more neural resources for decoding.

Eccentric fonts resound in the brain's response to novel stimuli. The aesthetics of distinctive typefaces can trigger activity in the brain regions associated with reward and pleasure, such as the ventral striatum and the prefrontal cortex. Enlisting unique fonts can strategically amplify interest and pique curiosity, beckoning consumers to explore further (Das et al. 2023).

Nostalgia takes the utilization of fonts a step further. Choice of typeface can imbue packaging with a personality or an essence of time and place. Script reminiscent of a bygone era speaks to the longing for authenticity, while futuristic lettering hints at innovation – a dance with the temporal that enlivens the history or potential within the brand.

As humans, our reading patterns have developed predominantly in a horizontal plane. Thus, when text departs from this norm – oriented vertically – the ease of reading stumbles. Such orientation disrupts the typical saccadic movement of the eye and challenges the natural flow of reading, subsequently increasing the cognitive load. This misstep can cause the brain to trip over the letters, obscuring the message in a tangle of effort.

When a simple statement or sentence spills over multiple lines, the cognitive load increases – effectiveness decreases.

Sentences on a package must be short, pithy, evoke emotion.

Circular positioning of text never works. When letters lean in different ways on different spaces of the package, the brain gives up.

Many fonts clutter, as each one cries for attention.

Translating the science of script into market speak reveals that the meticulous art of selecting a font for product packaging strikes a balance between invoking the consumer's emotion and preserving legibility. The engagement begins with the audience in mind, inviting them to follow the lead of well-chosen letterforms into a story – be it one of luxury, adventure, or comfort. The marketer's task is to match the font to the narrative's tempo, ensuring the brand's voice is clear and compelling.

Use GenAi to optimize the visual hierarchy of packaging design by selecting fonts that effectively differentiate between headlines, product names, and descriptive text. GenAi algorithms can analyze readability metrics to ensure that the chosen fonts are easily legible at various sizes.

An interesting application of GenAi could also be to generate custom fonts that are unique to the brand and resonate with consumers. This can involve combining existing font elements or creating entirely new typefaces tailored to the brand's personality and message. GenAi can also automatically check for font orientation and correct multiple orientations.

Magic of Touch: Texture and Haptics in Package Design

The phenomenon of tactile sensation extends far beyond the physical act of touch. It's a complex symphony orchestrated by the brain, where even visual cues can evoke the illusion of texture. As we encounter packages, our somatosensory cortex – the brain's tactile command center – fires in anticipation of textures we've learned to recognize visually. This neural prediction preemptively embraces the feel of a surface, engaging our senses and beckoning us closer, enticed by the promise of touch.

Research divulges how textures we see can invoke a tactile response even before our fingers graze a package's surface. The brain crafts a sensory preview based on prior experiences, allowing us to "feel" through sight. This ability transforms product packaging into a silent storyteller, whispering tales of the smooth, the gritty, the embossed – without a single touch.

Neuroaesthetics has shown that pleasing textures not only captivate but also retain our focus. When we observe textures considered aesthetically appealing, there is increased activity within the reward circuits of our brain, including the orbitofrontal cortex (Derke et al. 2023). Such textures command a longer gaze, weld a tighter bond between product and consumer, and may even command a premium for the pleasurable experience they propose.

It's fascinating to note how anticipated texture influences judgment about a product's quality. The engagement of the orbitofrontal cortex, a brain region associated with decision-making and subjective preferences, suggests that perceived texture quality can sway our estimation of the product's overall merit. This neurological bias for texture extends a powerful tool into the hands of packaging designers.

Packaging design that provides texture where effort is involved – holding, twisting, turning – is preferred, as it minimizes cognitive load to monitor success.

The tactile sensation is not solely a product of direct contact; it is also produced vicariously through our rich history of interactions with different materials. Neurons in the premotor and parietal cortices fire when we watch others performing actions. Similarly, when we witness textures, our brains can somatically simulate the feel, a process that can be enhanced or downplayed in packaging designs to either invite intimacy or exude exclusivity.

As we transition from the neurological underpinnings to practical application, consider the spectrum of commodification: luxury goods, with their silken linings and leather-bound cases, tell a story of sumptuous comfort. Meanwhile, rugged goods – that brandish a granular grip or a matte finish – convey durability and resilience. Both appeal to distinct psychological profiles, orchestrated through the visual whisper of texture that our brains are inclined to translate into somatic songs.

Packaging that has a ribbon around it and has soft inserts inside creates great appeal. This texture of a simple ribbon with the associated notion of gifting and special occasion creates powerful emotional appeal.

Employ GenAi to map tactile textures onto packaging design mockups, allowing designers to visualize how different textures interact with other design elements. GenAi algorithms can generate realistic renderings that accurately represent the tactile experience.

Utilize GenAi to design multisensory packaging experiences that combine tactile elements with visual and auditory cues. For example, incorporating embossed patterns or raised textures alongside vibrant colors and sounds can create a more immersive brand experience.

The Visual Vanishing Act: When Words Dissolve into Imagery

Words over an underlying image are generally ignored. The brain tries hard to erase the words to enjoy the underlying image.

Today's visual communication lies in the delicate dance between imagery and text. Neuroscientists have uncovered fascinating evidence of the competitive neural processing that occurs within the visual cortex. When words and complex backgrounds coexist, the brain prioritizes the processing of images due to our evolutionary wiring for rapid visual scene understanding. Hence, words superimposed on vibrant backgrounds can become the unsuspecting casualties of the brain's innate image preference.

Educational psychologist Richard Mayer's cognitive theory of multimedia learning elucidates how the brain processes visual and verbal information. His research posits that the brain deals with these dual channels separately yet concurrently, which should, in theory, allow for smooth integration. However, when the visual channel is overloaded, the verbal takes a backseat, making the text nearly imperceptible. Such cognitive congestion is an involuntary act of selective attention favoring the rich tapestry of shapes and hues over the austerity of alphanumeric text.

In design, the phenomenon of cognitive overshadowing, where one visual element suppresses another, must be dexterously navigated to ensure textual clarity (Yang et al. 2020). The design principle of signal-to-noise ratio – a term borrowed from the field of electronic communication – advises maintaining a high contrast between essential information (signal) and background (noise) to heighten message legibility.

When imagery contains hidden semantics – crop circles or words spelled in fields of grass, there is delight for the brain. Algorithms should utilize this cleverness of presentation.

It's not all science at the expense of aesthetics, though. The principle of figure-ground perception, derived from Gestalt psychology, advises that for text to triumph over a competing background, it must be clearly distinguishable as the figure against the ground of the image. Employing strategies such as sufficient contrast, text outlines, or strategic shading can flip the brain's switch to favor text readability.

Regarding front-of-pack strategy, science suggests leveraging negative space wisely. Negative space, or the area around and between subjects, is not merely an unoccupied canvas; it serves as the visual breather that allows text to emerge unscathed from battle with bustling backgrounds. The art of utilizing whitespace becomes not only an ode to minimalism but a tactical ploy to let words pop.

Utilize GenAi to ensure that words are not overlaid over complex texture and imagery as the brain will just erase the words to enjoy the image. Image generation algorithms should bear these in mind and know how best to position words in a complex image.

Simplicity in the Aisle: Rule of Threes in Packaging

In the busy store aisle, packaging with its vibrant colors and bold fonts competes for attention. Yet, amidst this visual cacophony lies a principle borrowed from cognitive neuroscience – the brain's predilection for the number three. The phenomenon, known as the *rule of threes,* observes that humans find it particularly effortless to grasp concepts and retain information divided into trios. This rule shapes narratives in literature, punctuates speeches with impact, and, relevant to our discourse, informs effective package design strategies.

Neuroscience demonstrates that patterns within the triadic framework are processed with remarkable fluency by the brain. The principle of numerosity, which addresses the brain's ability to recognize quantities without consciously counting, shows a preference for small quantities, typically three or fewer. This predisposition underlines an ease of cognitive processing that provides a mental shortcut to understanding – a neural fast track that packaging can leverage.

Statistical learning in cortex coherence relays how our brain recognizes and organizes information into patterns. The temporal lobe, involved in memory formation, exhibits an affinity for triadic groupings by building expectations based on frequencies and sequences of stimuli. Through this process, we effortlessly discern and remember patterns, an insight that packaging designers can exploit to make a product stand out.

Neural synchrony and triadic consonance further bolster the case for the rule of threes. Research into processing fluency finds that concepts presented in groups of three are not only easier to remember but also more satisfying to our penchant for patterns. A trio of imagery on packaging seems optimally structured for the constraints of human working memory, hence securing a podium placement in the attention stakes (Rapaille 2015).

As one selects images for the packaging trio, they tackle both homogeneity and variety with a nuanced brush. Symmetric designs draw from the brain's ancient fondness for balance, while a dash of asymmetry can tickle the orbitofrontal cortex with the unexpected. This delicate equilibrium between surprise and comfort, encapsulated in three parts, entices the consumer with its neuroaesthetic charm.

Utilize GenAi to structure packaging content according to the rule of threes, organizing information into three distinct spaces. Visually include no more than three big image groups.

Employ GenAi algorithms to group design elements, such as images, icons, or text blocks, into sets of three to create visual unity and coherence. GenAi can generate design variations that experiment with different arrangements and compositions to optimize visual impact.

Seeking Stability in an Uncertain World

Amid the aisles of a grocery store, a performance unfolds – products vie for a leading role in the consumer's daily drama. Beyond vibrant colors and bold fonts, an invisible character plays its part: gravity. Our brains possess an intuitive grasp of physics, honed by millions of years of evolution. The vestibular system, located in the inner ear, works closely with visual cues to maintain balance and spatial orientation, informing our perceptions about the stability of objects we view and manipulate. Hence, gravity's constant pull sets a fundamental context, shaping our expectations about how items should rest and behave on shelves.

The innate recognition of stability, or the lack thereof, feeds into cognitive biases that favor equilibrium. In the realm of product packaging, the cerebellum – a region of the brain adept at predicting the movement and stability of objects – together with the parietal cortex, springs into action when we interact with items on a shelf. These calculations are swift, subconscious, and speak to our desire for predictability and control, amplifying the attractiveness of a well-grounded design.

Packaging stability is thus allied with the brain's predictive mechanisms. The Fitts' law model, originating from studies in human-computer interaction, suggests that the time required to move to a target area (like picking up a product from a shelf) is a function of the target size and distance, as well as the stability of the object. Minimizing effort by ensuring easy retrieval – without risking a topple effect – resonates with this principle, reinforcing the consumer's preference for stable and accessible products (List & Kipp 2019).

The portrayal of stability in packaging also leverages the fundamentals of physical aesthetics. The brain appreciates symmetry, and when viewing structures, it tends to equate symmetrical shapes with stability. Therefore, a package that projects balance, through its visual and physical design, naturally appeals to these cognitive aesthetics, engaging the consumer with an ephemeral promise of order in their hands.

The optimally sized grip, the tactile feel of a smooth contour, and the satisfying click of a closing lid – all these features invoke the ancient brain machinery. The motor cortex lights up and hands reach out, drawn to the allure of a package that pledges a no-fuss retrieval and an equally slick return to its shelf perch.

Employ GenAi-driven algorithms to identify areas of the packaging that require reinforcement to improve stability. This could involve adding structural supports, cushioning materials, or reinforcement panels to critical areas prone to damage or deformation.

The Allure of Curves: Roundness Wins in Packaging

When it comes to consumer perception, sharp angles are not merely geometric shapes but are imbued with psychological implications. Neuroaesthetics, a field exploring the brain's responses to beauty and design, has dug deep into our neural responses to angular forms (Palumbo et al. 2015). Research suggests that angular shapes can activate the amygdala, associated with processing threats and fear (Larson et al. 2009). Evolutionarily, our brains have been wired to associate sharpness with danger as a survival mechanism – a sharp object signifying a potential threat. The implication for packaging is profound; sharp angles may subtly evoke a sense of caution rather than comfort, influencing the consumer's hand to pass over a product.

Curves counteract the amygdala's threat response, communicating safety and approachability. Rounded shapes and soft contours are seen as benign through the lens of our instinctual apparatus, inviting interaction and handling. These neural preferences are the silent sirens informing our perceptions long before conscious appreciation sets in. The curvature principle in design, therefore, not only pleases the eye but also aligns with the subconscious templates guiding our hand to reach for safety and comfort.

The preference for roundness has roots that intertwine with the brain's visual processing system. Rounded forms are processed in the inferotemporal cortex, a brain region involved in object recognition and visual perception. This fluency of processing translates to an ease of digestion. Simple, smoothly contoured packages don't just catch the eye; they converse with it in a language devoid of the sharp syntax that may cause a neural wince.

Our hands and fingers, dexterous marvels that they are, find rounded edges more comfortable to hold – a design consonance that goes back to the earliest tools our ancestors shaped for survival. Contemporary research in haptic perception supports this, indicating that our hands intuitively seek out shapes that promise no pain in the handling, a promise fulfilled by the voluptuary embrace of curves.

How does this translate to the supermarket shelf, you ask? Picture this: a row of bottles, jars, and boxes – all yearning for a shopper's touch. Those with angular edges might as well have donned tiny spikes, whispering "beware" to the subconscious sleuth inside every customer.

Utilize GenAi to optimize packaging designs for ergonomic comfort by incorporating round shapes that fit naturally in consumers' hands. GenAi algorithms can analyze grip patterns and hand anatomy to recommend shapes that enhance user experience and ease of handling.

Employ GenAi to ensure visual harmony between round shapes and other design elements on the packaging. GenAi-driven design tools can adjust proportions and alignments to create a cohesive and aesthetically pleasing overall design.

Imperfection = Authenticity: How Flaws Win Hearts in Packaging

Imperfection in product packaging imagery invites the brain's "beauty of imperfection." Neuroscience suggests that sensory appeal is far more nuanced than the mere appreciation of symmetry and cleanliness. The intricacies of our neural pathways are embroidered with a love for naturalness and authenticity, identifiable through the flaws we encounter. Perfection eludes human experience and hence meets with a meticulous, albeit subconscious, scrutiny of the artificial.

The orbitofrontal cortex, an area of the brain implicated in sensory integration and subjective aesthetic judgments, comes alive at the sight of authentic imagery. This neural activation underscores the visceral appeal of the "real" in our daily experiences. Images depicting a slightly asymmetrical apple or soup casually spilling over the edge of a bowl resonate with an innate recognition of the imperfect world we inhabit, grounding the product in the consumer's tangible reality.

More than visual stimuli, the psychological concept of Wabi-Sabi, an appreciation for beauty in the imperfect and transient nature of life, finds a home in the consumer's psyche. This ancient Japanese philosophy aligns with our tendency to value signs of use or wear, which suggest utility and reliability. These markers of imperfection can hint at a narrative, a product's journey before reaching our hands, and our brains are hardwired to savor such stories (Juniper 2011).

The decision to incorporate a hint of disarray into imagery does not denounce the pursuit of quality. Instead, it embodies a strategic embrace of the psychology of perception. A soup container with an image of a slightly messy bowl conjures a more impactful, genuine scene, liberating it from the sterility that might otherwise mute the chorus of neural affirmations.

Packaging imagery should reflect perfection that is "almost there," but with a few slip-ups. These few imperfections are translated by the brain to mean authentic – and therefore it seems very real and true to the consumer. GenAi algorithms must know how best to communicate the "little imperfections" that create the perception of authenticity.

Clear and Transparent: Seeing Is Believing in Product Packaging

The allure of transparency in packaging plays into the brain's preference for predictability and trust. Neurologically, transparency aligns with our preference for verifiable realities – we believe what we see. Visual transparency in product packaging affords our occipital lobes – the visual processing hub – an unobstructed view of the contents, marrying expectation with evidence, an essential concoction for trust formation.

Visual cues are given precedence in neural processing pathways, as they often provide the most immediate and concrete information about our environment. Dopamine, a key neurotransmitter associated with anticipatory pleasure and reward-based learning, spikes when the brain's prediction aligns with actual outcome. The less our prefrontal cortex has to work to predict what's inside the packaging, the more rewarding the viewing experience – hence, transparent packaging sings a siren song of simplicity and satisfaction to our neural circuitry.

The neural appeal of transparent packaging extends beyond the simple satisfaction of immediate visual confirmation. Research into aesthetic preference has shown a marked favoritism for object visibility (Simmonds et al. 2018), which falls under the broader psychological concept of ambient optic array – the richer the visual information, the better for our brain to map the environment. A packaging that offers a glimpse of its contents is like a window in an otherwise opaque wall of consumer choice (Sutherland 2020).

The hippocampus, ever involved in memory formation, binds together visual information with previous experiences. The ability to see a product within its packaging can trigger memories. The crystal-clear view of shampoo might not just signal its quantity but evoke tactile and olfactory flashbacks of previous encounters with the product. It's a neurological nostalgia trip facilitated by a panel of plastic or glass.

The choice to display the olive oil's slow descent or the perfume's vibrant hue isn't mere aesthetics; it is a strategic seduction of sight. When we peer through a transparent food container, we're not just scanning for spoilage or quantity; we're engaging in an unstated pact with the product: You show me your true colors and I'll bring you home.

Humor naturally bubbles up in the interplay between brain and transparency. Picture a bottle of ketchup that reads, "Squeeze me! I'm not shy," as it brazenly showcases its viscous red contents. It's a playful wink that not only leverages our brain's love for visibility but also incites a chuckle – engaging the brain's reward center and ensuring a sticky memory of the brand, long after the fries are gone.

Utilize GenAi to create virtual prototypes of transparent packaging designs, allowing marketers to visualize how the product will appear in different packaging configurations. GenAi algorithms can generate realistic renderings that accurately depict the product inside the packaging.

Employ GenAi-driven image recognition algorithms to identify key product features and highlights, ensuring they are prominently displayed in transparent packaging. GenAi can analyze product images and recommend positioning strategies to maximize visual impact.

Clicks and Snaps: Auditory Cues in Packaging

Auditory processing is an intricate part of the brain's sensory experience, often occurring without conscious thought. The auditory cortex, bilaterally spread across the temporal lobes, is tasked with decoding and assigning meaning to the diverse sounds we encounter. This power of sound extends to the world of packaging, where a simple sound can elevate a product from mundane to memorable, embedding the experience within our neural networks.

The brain's propensity for sound is more than a mere fascination; it is rooted in survival. Sound has historically signaled both danger and opportunity, key to our ancestors' survival. Modern neuroscience appreciates this soundscape, showing that positive auditory feedback, such as a click or a snap, can stimulate the reward pathways, releasing dopamine – the feel-good neurotransmitter (Weis et al. 2013). Beyond the reward system, the role of auditory feedback in task confirmation is crucial. The superior temporal gyrus, active in auditory processing, assists in pattern recognition and the anticipation of sound sequences. This neurological function underpins the pleasure derived from a sound that confirms an action, such as the definitive click of a container lid, indicating a secure close.

As our exploration treads into the alleys of packaging design, the theatricality of sound takes center stage. The auditory dimensions that greet us after the visual cues are vital players in the user experience. It's not simply about securing the contents; it's about performing a duet with the consumer's expectations, where the satisfying snap of a make-up compact signals a case safely sealed, composed by design to resound with cognitive approvals.

Employ GenAi-driven techniques to integrate auditory cues into packaging design – whether it is the click or a pop or a satisfying thud. GenAi can recommend suitable materials or technologies that produce desired sound effects when activated, such as clicking buttons or snapping closures.

The Pop-Out Paradigm – Enabling Cognitive Fluency

On crowded shelves, certain designs capture our gaze with the force of a lighthouse, standing boldly against the monochrome of the mundane. This phenomenon taps into the brain's visual search mechanisms, where a feature that markedly deviates from its peers "pops out," demanding our attention with minimal cognitive effort (Kerzel & Schönhammer 2013). When a hue interrupts the grayscale of expectation, this "purple" against a "black and white" landscape is instantaneously processed by the visual cortex – a neural beacon in the fog of similarity.

Contrast acts as a visual clarion call to our attention system, leveraging the brain's predisposition for detecting outliers. Neuroimaging studies have highlighted the increased activation in areas such as the lateral occipital complex when contrast is present, aligning with our innate interest in distinctiveness over homogeneity. This neurological preference for contrast enables well-designed packaging to leap from shelves into the consumer's awareness with the subtlety of a pyrotechnic display.

But it's not just color that plays this cunning game of hide and seek with the brain's filters; other sensory features like size, shape, or motion can also contribute to this "pop-out" effect. A package might achieve prominence through an unexpected texture or an atypical form, engaging the brain's attentional networks and drawing the eye with the gravitational pull of the novel.

Science becomes practice when designers employ these principles to make products stand out on the crowded canvas of commerce. Clad in colors that contrast or adorned with elements that disrupt visual expectancy, these packages become the mavericks in an assembly line of conformity, proudly proclaiming, "Look at me!"

In the lighter vein of application, envision a shopper's delight when a sudden burst of color amidst a sea of beige beckons them like the Pied Piper's tune. It's an unspoken conversation in the dialect of visuals, where the shopper's brain applauds the break from monotony with every neuronic cheer.

Utilize GenAi-driven design tools to incorporate pop-out elements, such as embossed textures, raised graphics, or holographic effects, into packaging design. GenAi algorithms can generate design concepts that add tactile and visual interest, enhancing consumer engagement and interaction.

Employ GenAi to optimize visual contrast and readability in packaging design by strategically placing pop-out elements against background colors or patterns. GenAi-driven layout optimization techniques can enhance visual hierarchy and draw attention to key product features or messaging.

The Allure of Illusion – Motion in the Mind

Optical illusions are not mere parlor tricks; they are a cerebral escapade, a fascinating manipulation of the brain's perceptual machinery. These mind-bending displays ensnare the occipital lobe, which houses the brain's primary visual cortex. They exploit the brain's tendency to infer depth and perspective where none exists, creating an irresistible allure as the brain juggles with reconciling conflicting cues.

These visual riddles activate numerous cortical areas, including the dorsolateral prefrontal cortex, which is involved in high-order cognition and resolving ambiguity (von Gal et al. 2023). By engaging such complex cognitive processing, illusions woven into packaging design can transform the product into an enigmatic figure that refuses to be ignored.

Holograms and printed illusions on packaging perform an entrancing waltz in front of our eyes, their every deceptive step meticulously choreographed to enrapture the onlooker (Rauschnabel 2021). The brain delights in these puzzles, rewarding the viewer with a dose of neurotransmitters associated with pleasure and surprise as they decipher the visual enigma.

Employ GenAi-driven design tools to seamlessly integrate optical illusions into packaging design. GenAi algorithms can generate design concepts that leverage techniques such as trompe-l'oeil, Moiré patterns, or perspective tricks to create visually captivating effects.

Leverage GenAi-driven techniques to create illusions of depth and dimensionality that make packaging appear larger, smaller, or more spacious than it actually is. GenAi algorithms can generate designs that use shadowing, shading, or overlapping elements to create immersive visual effects.

The Mirror Effect – Beautiful Me

The mirror effect in packaging design capitalizes on our brain's extraordinary response to self-recognition. A reflective surface on a package can activate the right parietal cortex, the key player in the perception of one's own body in space. Moreover, it stimulates the neural networks associated with self-reflection and personal identity, including the insular cortex and the anterior cingulate.

The act of recognizing oneself in a reflective surface has been associated with an increase in self-referential processing, leading to the subjective valuation system in the ventromedial prefrontal cortex becoming more active. This phenomenon suggests that products with mirror-like packaging might not only capture the eye but also resonate on a more personal level.

Transforming this fascinating neural response into a tangible benefit, packaging that incorporates reflective elements reaches out to consumers with the subtle flattery of their own reflection. It's as if the product whispers, "You are part of this too," forging a connection between the consumer's self-image and the item at hand.

Utilize GenAi to create virtual prototypes of packaging designs featuring mirror effects or reflective elements. GenAi algorithms can generate realistic renderings that accurately depict how light interacts with reflective surfaces, allowing marketers to visualize the final packaging design.

This magnetic attraction to our own reflection emanates from the physiological mechanisms of the right parietal cortex, a region of spatial awareness, stirred into action as we catch a glimpse of ourselves on a shiny package surface. The encounter with a mirror-like package invokes an intimate, personal touch, as the brain's networks conflate the self-image with the product. It becomes more than a product – it's a reflection of the consumer.

It is therefore important to never present chopped off hands and limbs as part of imagery on a package. A natural aversion reaction to mutilation may make the consumer walk away from the package.

Beyond the allure of self-recognition, reflective packaging also plays tricks with ambient light, creating a dynamic dance that keeps the brain engaged. Light sparkles in the consumer's eyes, and the superior colliculus, tasked with orienting reactions to visual stimuli, catapults their focus onto the radiant packaging. It's a visual conversation amplified by the brain's deep-rooted celebration of light – a feature that has guided our ancestors through the genesis of cognition.

The strategic incorporation of reflective properties into packaging design does more than make the product pop; it wields the power to enchant. This encounter is not soon forgotten, as the brain's reward system makes a mental note of the pleasurable experience (Mason et al. 2017).

Employ GenAi-driven material selection tools to identify suitable materials and coatings that create mirror effects or reflective surfaces on packaging. GenAi algorithms can recommend options such as metallic foils, holographic films, or glossy finishes that enhance reflectivity and visual appeal.

Expect the Unexpected – Packages That Are Novel

The pursuit of the novel drives much of human exploration, and this is no different in the landscape of product packaging. The brain's novelty detection system, centered in the substantia nigra/ventral tegmental area (SN/VTA), triggers a pleasant burst of dopamine when consumers encounter something new and unexpected (Krebs et al. 2011). This neurochemical reward for novelty ensures that a product differing in form, color, or method of interaction from its peers crisply imprints itself within the consumer's memory.

Whether it's through a new tactile texture, an unusual shape that defies convention, or an innovative method of revealing the contents, engaging the brain's love of novelty is akin to a psychological embrace that lingers. The newness sparks a conversation within the brain's default mode network as it processes the unfamiliarity, integrating it into the self-referential narrative.

By designing packaging that steps away from the expected, brands can break the autopilot of consumer behavior, instigating a refreshing pause in the otherwise uninterrupted flow of shopping habits. Newness can be as subtle as a surprising pop of color on a label or as dramatic as a transformative packaging concept that shifts form and function in unexpected ways.

Correct the Error – Package with an Apparent Error

The brand French Connection, with its logo "FCUK" figured it right. The brain races to correct the error, smiles, and pays attention to know more. A simple "apparent" error, but wins the most precious commodity in the aisle – consumer attention.

The retail landscape teems with a myriad of products, yet certain packages prompt a neurological double-take – a smirking nod from the brain as it delights in an intentional blunder. This psychological tug-of-war hinges on the error-related negativity (ERN), a signal within the anterior cingulate cortex (ACC) that erupts when humans detect errors (Holroyd et al. 2004). Product packaging employing clever plays on words or graphic faux pas harnesses ERN, granting the consumer a triumphant moment of cognitive correction.

Delightfully designed errors are not fleeting; they impress upon the consumer's declarative memory, nestled within the hippocampus. Ingenious nomenclature akin to the playful "FCUK" brand sparks a collaborative dance between memory encoding and emotional arousal, courtesy of the amygdala's penchant for novel and emotion-laden stimuli. The reward for this mental gymnastics is a hearty release of dopamine from the substantia nigra, as the brain celebrates its own shrewdness.

The blend between error and rectification must be a smooth cocktail, neither too cryptic to decipher nor too transparent to unveil – aiming for a gratifying hiccup over a disconcerting stumble. Such designs flirt with the boundaries of expectancy violation, a principle that surmises our brain's attraction to mildly unexpected occurrences while maintaining overall coherence.

This subtly orchestrated error begets a cognitive tickle, a synaptic giggle elicited by the acknowledgment of the clever ruse. The shopper, upon discerning the nuanced faux pas, experiences the cerebral equivalent of an "Aha!" epiphany – an electrophysiological signature of realization marked by a wave of P300 activity in the parietal lobe (Knight 2000).

From a marketing strategy standpoint, the ideal correctable error might lurk within a typographical jostle or an ambiguous tagline. It's an engaging puzzle encapsulated within the packaging that says, "Solve me, and bask in the delight of your keen perception." With subtlety and wit, packages that whisper these brainy riddles become the heroes of the shopping aisle (Parker 2022).

The playful imperfection embedded in the design is not merely visually memorable; it converses with the brain's love for resolving incongruence. The somatosensory cortex plays along as buyers touch and scrutinize the packaging, their tactile exploration rich with anticipation for the delightful inconsistency presented. The brand "FCUK" clearly knew what it was going for.

Utilize GenAi to generate custom errors or design anomalies that evoke a sense of playfulness and creativity. GenAi algorithms can analyze brand attributes, consumer preferences, and cultural trends to recommend error concepts that resonate with the target audience. GenAi algorithms can generate designs that incorporate errors in typography, graphics, or layout elements in a visually appealing and cohesive manner.

Partly Covered and Partly Revealed – Seduction in the Aisle

The allure of ambiguity in product packaging leverages the brain's desire for narrative and mystery. This cerebral seduction is rooted in the concept of perceptual curiosity, where partial information tantalizes the prefrontal cortex, the brain's epicenter for problem-solving and inference (Van Lieshout et al. 2018). When packaging artfully obscures a portion of its contents, it sparks a visual flirtation, a suggestive peekaboo that beckons the brain's attention and engagement.

The tease of revelation is no trivial affair; it draws on the power of predictive coding, a theory suggesting that the brain is a hypothesis-testing organ that becomes particularly active when confronted with ambiguity. As consumers are lured by the elegantly veiled product, the inferotemporal cortex, which specializes in visual object recognition, buzzes with the anticipation of discovery.

Crafting packaging with an aura of mystery is akin to composing a sonnet for the senses. It's a deliberate dance between the seen and unseen, deftly choreographed to titillate the superior temporal sulcus, attuned to both motion and the nuances of visual storytelling.

An elegantly executed package of this kind plays coy with visual cues, inciting a cerebral game of hide and seek. As this dance of ambiguity unfolds, the brain's visual cortex and imaginal realms are coaxed into attentive wakefulness. It entices one to ponder, "What lies beneath the lavish folds of packaging?" The insula, associated with emotional responses and the feeling of anticipation, fuels the desire to explore and uncover.

This murmur of mystery does not culminate at first glance. It invites lingering looks and tactile engagement, enticing fingers to trace the outlines obscured by layers of mystique. Every peel and reveal become chapters in an unfolding romance novel between consumer and product.

Employ GenAi-driven design tools to incorporate hidden elements or cryptic symbols into packaging design. GenAi algorithms can generate designs that feature subtle clues, riddles, or puzzles that encourage consumers to unravel the mystery and engage with the packaging on a deeper level.

Tall Packages Win

In the battle for retail, where each product competes for a leading role in consumers' carts, some become consumer favorites by virtue of stature alone. The allure of height in product packaging stands tall as a monument within the consumer's neural pathways. Size perception and its implications for value are deeply rooted in our parietal lobes, the somatosensory processing centers (Hoba et al. 2022). The vertical advantage of packaging is a whisper to our primitive instincts – taller is mightier, commanding respect and attention through its upright disposition.

Height in packaging is not a mere trick of the eye but a calculated play on the brain's innate associations. When juxtaposed against shorter, albeit equivalent volumes, the lofty package triggers a perceptual magnification within the intraparietal sulcus, swaying the scales of perceived worthiness. It's as if the product, through its aristocratic height, announces a silent decree of superior quality from the shelves.

Every inch ascended by packaging in its design is a step toward cognitive distinction. The taller profile of a product breaks the monotonous horizon of homogeneously horizontal offerings, catching the eye of the beholder by gently nudging the brain's superior colliculus – the gatekeeper of our visual and spatial attention.

Utilize GenAi to optimize packaging height for enhanced shelf visibility and standout appeal. GenAi-driven design techniques can generate packaging designs that leverage height to draw consumers' attention and distinguish products from competitors on crowded retail shelves.

The Illusion of More Substantive with Weight

The weight of an object draws a direct line to its value etched within the consumer's brain. The grounding effect of a product's heft on perceived worth is a symphony orchestrated by the somatosensory cortices, the neural maestros of tactile sensation and weight perception. When an item bears with it a substantial mass, the brain ignites circuits of appraisal, as historically, weight has been a touchstone for material quality and importance.

The sensation of heft is more than a muscular judgment; it is an encounter with the vestibular system's estimates of material wealth. Heavier objects elicit a more pronounced activation in regions associated with value – such as the orbitofrontal cortex – intimating that in neural currency, weight translates to worth (Jostmann et al. 2009). It's a kinesthetic whisper that resounds with the gravitas of affluence.

The tactile dance with a weightier item is an interplay between our hands' mechanoreceptors and the brain's reward pathways. Heft in hand signals distinction to the pleasure centers, crafting a tactile intimacy that reinforces the consumer's choice as one of substance – literally.

The thick glass of a perfume bottle or the dense board of a luxury item's boxing are no accidental choices; they are tactile sonnets serenading the shopper, where each gram sings a verse of value. It's the brain's ballad of the profound, played on the strings of product packaging with deliberate measure – where each ounce offers its symphony of significance.

The charm of heft in packaging is an artful blurring of lines between physical mass and psychological massiveness, where what is grasped in the palm is simultaneously embraced by the palette of perception – a seamless synthesis of the material and the mental.

Employ GenAi-driven simulation tools to digitally model packaging designs and simulate their physical weight and feel. GenAi algorithms can generate virtual prototypes that accurately depict the perceived heft and tactile qualities of packaging materials, allowing marketers to evaluate different design options.

Utilize GenAi to analyze different packaging materials and their perceived weight, durability, and premium-ness. GenAi algorithms can recommend materials such as thick cardboard, rigid plastics, or metal accents that convey a sense of heft and solidity to consumers.

Left versus Right: Words and Images

It is not just true in politics, it is true in the processing of the package as well.

Hemispherical specializations enable slightly different kinds of processing in the brain. Images on the left, received by the right frontal are processed effectively. Semantics – words and numbers presented in the right visual field are processed well by the left frontal. Switching them slightly increases the cognitive load, and may lead to decreased effectiveness of the presented package overall.

Employ GenAi-driven simulation tools to consistently ensure that images are prominent on the left of the package and words and numbers are prominent on the right side of the package.

If a brand is written as a word, it is still processed as an image.

Utilize GenAi to verify the proper positioning of the words and imagery in a package.

In this chapter we have presented the core neurodesign elements for packaging that captivates the human mind. We have shown how GenAi algorithms can embed these principles as imagery, words, shapes, sizes, textures, and fonts for packaging.

CHAPTER 14

fragranceGPT and flavorGPT – neuroAi-Powered Design of Fragrances and Flavors

T he typical scenario today.

Brand leader invites one of the four big fragrance providers to create something new for the Gen Z consumer.

Consumer is a young, hip woman in financial services in Manhattan.

Fragrance company comes back after three months and recommends top notes of patchouli, lime, and jasmine, middle notes of rose and orange, and base notes of cedar and sage.

When asked how they leaped from GenZ woman in Manhattan to patchouli, lime and jasmine, and if they could provide a transparent path and reasoning, the fragrance company responds, "Trust us, we've done this for years. We have perfumers in Paris who do this for a living. (Don't you dare question us.)"

When asked if we can have the formula for the fragrance, they say, "Trust us, you don't want it, just license it from us. (Haha, we won't give it to you.)"

The same drama happens with flavors.

The question in everyone's mind that nobody has the courage to articulate, "Excuse me, but why exactly should we trust you? You have shown us nothing about your methods and have not shared your reasons."

The Gustatory Complex – What Makes Life Worth Living

For many people, the hardest part of getting COVID was not the illness but the loss of taste and smell. Our senses guide us through a world filled with desires. Seeing, smelling, or hearing something can instantly remind us of past experiences because of how our brains store memories. When we talk about why certain smells make us feel a certain way, it's because specific parts of our brain, like the insular cortex, help us perceive and process these sensory details.

The sense of smell is directly connected to parts of the brain that handle our emotions and memories, such as the amygdala and hippocampus. When we catch a whiff of a particular scent, tiny sensors in our nose send signals directly to these emotional and memory centers. This connection can have a powerful effect: A single smell can flood us with feelings or transport us back to a moment from our past.

In this chapter we will look at the power of scents and fragrances, and how their design can be effectively guided by neuroscientific principles and the memory structures we encountered in earlier chapters. Finally, as a culmination, we will look at how GenAi can be used to design tantalizing fragrances.

The link between smell and memory is not just fascinating; it's a powerful tool. This relationship offers a treasure trove of opportunities for businesses to build stronger brand connections. Because our sense of smell and memories are so closely knit, smells can quickly bring back emotions or moments.

Our noses pick up countless scents, each capable of triggering vivid memories. What makes smell unique is its direct line to the brain's centers for emotion and memory. Unlike the other senses, smell is unique in how it bypasses the thalamus and heads straight for the brain's emotional and memory centers (Rolls 2019). With every breath, we have the ability to relive past moments. Marketers can use this shortcut to our feelings by linking their brands with specific scents. By doing so, they create lasting impressions that are stronger than any slogan or tune could ever be.

Remember, a brand's chosen scent is more than just a smell – it's a beacon for making memories. The sense of smell possesses a unique character due to its direct link to the hippocampus, a central area for memory consolidation. This means that scents have a direct neurological link to memory storage, making it possible for a brand's distinct scent to firmly lodge itself in our recollection. By crafting and maintaining a unique smell, brands can become unforgettable.

Neuroscience has consistently demonstrated that our sense of smell can trigger profound emotional responses, often more so than other senses. A well-chosen fragrance can tap deep into the psyche, evoking a spectrum of feelings that resonate with a brand's narrative. This principle reveals a potent strategy for marketers: infusing products with smells that resonate on an emotional frequency, creating a resonance that consumers can feel deeply and personally. It's not just selling a product – it's selling a feeling. The scent is personal and subjectively experienced. This offers a unique opportunity for brand differentiation. Individual preferences for certain scents can affect the perceived value and desirability of products. Crafting a unique signature scent for a brand means forging a personal connection with the consumer, using aroma as an invisible thread that draws them toward the brand narrative and keeps them enmeshed within it.

The marketplace is a vast space of competing stimuli, but incorporating scents lends marketers an invisible tool to stand out.

Nonconscious Roadmap into the Consumer's Fragrance and Flavor Preferences

In the previous section we detailed how fragrances are fundamentally connected to emotions and memories. Fragrances are capable of reviving long-forgotten memories or feelings with the faintest sniff (De Brujin & Bender 2018). It is therefore a potent tool in the evocation of desire.

Creating desire using fragrance requires an understanding of the nonconscious that ties customers to your products. By crafting experiences, desire is created one aromatic note at a time, guiding consumers through an olfactory narrative. We have already introduced the concept of memory structures and detailed how they allow the probing of resonances in the nonconscious mind. In this section we will revisit the idea of these memory structures and explain how they connect to the development of fragrances.

Identifying the perfect aroma for a brand involves more than mere selection – it's an artful blend of science and sensory intuition. The olfactory signature must be aligned with the brand's image and the consumer's psyche. A complex emotional identity can be mirrored in the fragrances chosen for a brand – each note targeting a specific mood or sentiment. Narrative transportation theory suggests that well-crafted stories can immerse individuals, leading to heightened emotional involvement and persuasion. Each product becomes more than its function; odors orchestrate the mood, setting, and consumer experience. Marketers need to think of themselves as olfactory artists selecting from a palette of scents to evoke the desired emotional spectrum within their target audience.

Recall the following key memory structures:

- **Spatial memory** – relating to how we visualize the world through places, occasions, directions, visuals, and scenes
- **Temporal memory** – relating to seasons, times of day
- **Procedural memory** – rooted in daily rituals habits and routines
- **Emotional memory** – deeply connected emotional responses
- **Semantic memory** – archived knowledge of the world
- **Episodic memory** – transformed personal experiences and connections

We have already seen how fragrances connect in a fundamental way to emotional memory. However, fragrances connect to all of the memory structures.

These associations can be leveraged to identify notes and smells that connect to the nonconscious mind.

The principles of classical conditioning explain much about our scent-driven choices. In the early twentieth century, Pavlov demonstrated that dogs could learn to associate a bell with food, leading to salivation at the sound alone. Within the human marketplace, scents play the role of the bell, with products as the proverbial feast. An aroma repeated alongside a brand becomes intertwined in the consumer's brain, triggering emotions and influencing behavior.

Scents do more than just remind. They shape emotions – tailoring our mood with a mere whiff (Seubert et al. 2009). The scent of lavender isn't simply relaxing by the merit of the culture (emotional memory); it's scientifically acknowledged for easing stress reactions within the nervous system. Clever marketers use such ambient fragrances to create a tranquil shopping environment, skillfully weaving calm into the very air customers breathe. The scent of freshly baked cookies, for instance, doesn't just tickle the nostrils; it triggers a cascade of dopamine, the pleasure neurotransmitter, which encourages us to pursue such gratifying experiences again (emotional memory).

In each case, the culture, the country, the demography and the economic strata of the consumer determine the particular kinds of flowers, fruits, fragrances, foods, beverages, and flavors that will be present. So these memory structures become specialized for every single consumer demography.

Fragrances and Flavors in the Places We Have Lived and Traveled and Been

The fragrance of burning candles may evoke the tranquility of meditative spaces (spatial memory), or a particular cologne might bring back the rush of a first date's nervous excitement (emotional memory). These connections are created and reinforced over time, with repeated exposure strengthening the neural associations.

Layering scents to complement a brand's identity is a deliberate strategy of layering aromatic notes designed to resonate with the consumer's subconscious brand image. Brew a coffee scent in a bookstore (spatial memory), and customers don't just see volumes of print; they feel the comfort of mornings (procedural memory), cradling warm mugs with a novel in hand. However, subtlety is paramount. The intensity and profile of the scent should be just enough to enrich without overwhelming.

Understanding the places the consumer frequents gives us clues to the flowers, fruits, foods, and fragrances they have connected with, and created strong associations with. These fragrances and flavors have been successfully integrated into their nonconscious.

Utilize GenAi to analyze spatial contexts and environmental cues associated with target consumer segments. GenAi algorithms can generate insights into spatial preferences, sensory perceptions, and spatial memory structures, informing fragrance development strategies tailored to specific spatial contexts.

Fragrances and Flavors Evoked by the Passing of Time

The seasons carry their signature scents, curated not just by nature but also by culture.

Neuroscience informs us that certain fragrances are seasonally encoded within our cognitive processes, inducing particular emotional states (Bentley et al. 2023). The amygdala and hippocampus work in tandem, reacting to the familiar spices of autumn or the fresh citrus of summer, and linking these scents with seasonal behaviors and traditions.

This scent-season synchrony is no coincidence; it emerges through years of ritualistic associations from celebratory gatherings. Naturally, the marketing of seasonal promotions is well poised to exploit these encoded olfactory preferences. Festive experiences are often accompanied by distinct scents – pine during Christmas, barbecued delights on the Fourth of July, or floral fragrances heralding spring festivals. These associations occur due to the pattern encoding within the hippocampus, which links scents with specific temporal events. It is these recurring patterns that create predictive cues for future behavior. In marketing, the aromas of festivities can be strategically deployed to harness the predictive and associative qualities of scent. A waft of cinnamon might not just invite thoughts of pumpkin pies but can subconsciously cue the consumer's holiday shopping behaviors, ingrained from perennial rituals of festivity-fueled commerce.

Employ GenAi-driven customization strategies to tailor fragrances based on historical periods, time of day, or seasonal events. GenAi algorithms can recommend scent variations or customization options that align with consumers' temporal preferences, enhancing the temporal relevance and emotional resonance of the fragrance experience.

Fragrances and Flavors in the Rituals and Routines of Our Daily Lives

Our actions are often invisibly guided by the scents that pervade our environment. The basal ganglia, a group of nuclei deeply entrenched within the cerebral hemispheres, play a pivotal role in the formation and automation of habits (Lipton et al. 2019). This part of the brain works tirelessly behind the scenes, reinforcing the neural loops that associate specific olfactory cues with our everyday rituals. The aroma of morning coffee, for instance, is not merely a sensory pleasure; it becomes a neural Post-it note, reminding us of the routine that accompanies the first sip – kickstarting our day.

Our olfactory system marks the passage of time through scent-infused landmarks. The entorhinal cortex, the odorous map of our temporal lobe, aids in this process by facilitating olfactory memory and navigation. It is in this neurological waystation that the scent of a crackling fireplace becomes an evening's closing ceremony. Like the tolling of a clock tower, aromas punctuate our day, providing an olfactive rhythm to which we subconsciously march.

Scents and fragrance notes are deeply connected through all memory structures, and clever use of such associations can place the consumer in a conducive state of mind, transporting them to specific spaces, seasons, and emotional associations. Several of these can also be layered and combined. Neuroscience suggests the olfactory system processes these various notes differently, etching auras of emotions into the cortex.

Utilize GenAi to analyze procedural contexts and daily routines associated with target consumer segments. GenAi algorithms can generate insights into procedural preferences, sensory perceptions, and procedural memory structures, informing fragrance development strategies tailored to specific procedural contexts.

It's not just familiar scents that hold power – new and innovative ones can carve their own neural pathways in consumers' memories. The brain's plasticity allows for the creation and reinforcement of new associations, even in adulthood (Erickson et al. 2013). Thus, a novel fragrance can become desired through its very novelty.

A fragrance is composed of various notes across various layers (something we will see in detail shortly). The top notes might sing of innovation, while the base notes hum with reliability. Citrus can convey cleanliness and efficiency; vanilla can impart comfort and warmth, while peppermint can energize and refresh.

The art of fragrance creation that appeals to the nonconscious is not just about the opening top notes, but it is about telling a story through the entire duration of the fragrance, culminating in the "after-sniff," an echo of the aroma that reverberates through the consumer's memory long after the initial encounter. Just as a catchy tune or a poignant line from a film can linger in one's thoughts, so too can a masterfully crafted fragrance maintain a presence in the mind, continuing to influence brand perception and consumer behavior.

GenAi can analyze consumer data to predict scent preferences based on demographic information, purchasing history, and social media interactions. GenAi can analyze competitors' fragrance offerings to identify gaps in the market and opportunities for innovation. GenAi algorithms can additionally identify emerging fragrance trends by analyzing social media posts, online reviews, and industry publications.

Fragrances and Flavors in Charged Emotional Moments

The smell of roses and marigolds can usually be associated with the elaborate rituals of a multi-day Indian wedding with the scent of curries and cumin and cardamom wafting through the air.

Vanilla and cake might connect deeply to a festive birthday party.

The smell of damp soil might connect to relief, and a refreshing and much needed rain shower on a hot summer day.

Emotions are connected to fragrances and flavors deeply. Understanding and extracting the emotional moments in a consumer's life can reveal the fragrances and foods that became part of context and background on that day.

GenAi can analyze consumer's emotional moments and extract the fragrances and flavors that formed the foundation and backdrop for those moments.

Creating a Distinctive Scent: Neurodesign Using Memory Structures

In this section, we will look at how the intrinsic associations of smells with memory structures can be leveraged in the design of fragrances with GenAi.

Artificial intelligence brings a systematic approach to understanding and innovating in this universe of scent. Using GenAi and large language learning models (LLMs), the traditional process of scent design can be revolutionized. Backed by advanced technology, GenAi not only helps in creating unique, captivating scents, but also ensures cost-effectiveness and efficacy against rival fragrances.

Perfume crafting has traditionally been the domain of skilled, selective humans who use personal experience and detailed briefs to create exclusive fragrances. The key breakthroughs in technology allow AI to replicate a significant part of this process. GenAi and LLMs can take a given brief and come up with a multitude of fragrance versions. Every detail of these alternative scents, including their full chemical composition, safety data, and associated certifications, can be generated autonomously.

Similarly, we often look to chefs or grandmothers for the best in the culinary world. But LLMs can learn from millions of recipes, studying and modeling the pairing of ingredients and their preparations to generate unique and imaginative culinary creations.

Fragrances are ubiquitous in consumer products, from shampoos and soaps to toothpastes and detergents. Brands are looking to formulate unique, cost effective, well-balanced fragrances that align with their brand and will be liked by their consumers. Additionally, the new fragrance has to perform well against key competitors.

However, it is possible, using neuro-gen AI algorithms, with the same input that a brand provides to a perfume house, to set up transparent end-to-end algorithms that can compose several fragrance options. These fragrances can then be realized from concept to sample and eventual manufacture.

In order to better understand some of the details that follow, it will help to describe why formulating fragrances is a challenge. Typically, fragrances are described in terms of their top, middle, and base notes. A note generally corresponds to a particular scent, such as vanilla or musk.

The top notes are composed of compounds that are most volatile and therefore evaporate first. These are the notes that can be smelt first. As time goes by,

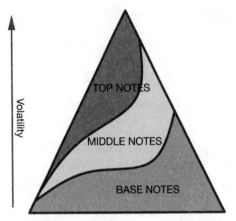

FIGURE 14.1   A diagrammatic representation of a well-balanced fragrance and its temporal structure.

the slightly less-volatile compounds come to dominate the fragrance, and these are the middle notes. And the base notes are the last lingering compounds of the fragrance.

The challenge is to construct a fragrance that will be well balanced across time, with properly blended top/middle/base notes.

The scent design process starts with the identification of "seed notes" that resonate clearly with the brand's target consumers. These seed notes are the scents and smells that connect deeply to both the nonconscious memory structures of the consumers in the context of the associations a brand might want to pursue. The seed notes are extracted through GenAi models that are guided by frameworks to identify and extract notes and scents that would be familiar within the target context to consumers – specific to geography, age, gender. This is where the memory structures come in handy. They allow us to find the fragrance and flavors that are associated with these memory structures. The places we have been, the changing of the seasons of our lives, and the rituals and routines of our daily lives have nonconsciously created firm and deep associations with fragrance, flavors, and ingredients.

These identified notes are then used by the GenAi platform to curate and construct a host of possible fragrance combinations. An AI algorithm then ranks these potential compositions. Several factors, including the brand's profile, demographic target, and market trends, guide the decision-making process, allowing the AI system to make the most balanced, customer-appealing fragrance. Much like LLMs are traditionally fed text and natural language, in this case such models are fed a large database of existing fine fragrances designed by master perfumers from across the world, to learn implicit rules of composition. These models learn an internal representation of olfactive rules of composition – something a master perfumer spends decades honing. Equivalency in flavor space is a million and a half recipes created and curated by master chefs.

However, simply throwing together ingredients is not enough. Clearly the fragrance must be well balanced, cost-effective, and do a good job in the desired medium – be it shampoo, soaps, detergents, creams, etc. Furthermore, these fragrances need to meet extremely stringent criteria for safety. Therefore, a crucial aspect of this innovative technology-driven approach is the maintenance of safety standards and compliance with IFRA (International Fragrance Association) regulations. The GenAi system meticulously adheres to IFRA's standards to ensure the new fragrance is safe for consumer use. The resulting composition is not only uniquely crafted but also accompanied by all the essential safety data. Validating perfumes using AI ensures complete transparency in scent creation, further enhancing the trust in the brands. At the end of this AI-driven rigorous process, the brand fully owns the fragrance composition designed specifically for its consumers.

The Olfactive Metric Space

The notion of distance, or a metric, is very important in this space. Here one might choose to create a complex metric that determines the distance between two fragrances in this space. One could also treat each fragrance as a sentence consisting of "note words" and represent each note as an embedding in a higher dimensional space. There exist a few approaches to create metrics in the world of fragrance and flavor. Defining metrics in this space allows one to ask the following key questions:

1. How far is the fragrance I have designed from the fragrance of the current product?
2. How far is the fragrance I have designed from the fragrance of the competitor?
3. How far is the fragrance I have designed from the best sellers in the marketplace?

Quantitative answers to these questions allow us to determine the key question every fragrance maker desires to answer: How unique is my fragrance?

The Olfactive Space – Fragrance and Flavor Design Using GenAi

When it comes to flavors, there is a world similar to top, middle, and base notes, except we think in terms of aromatics, spices, protein, greens, fats and other categories of ingredients. Much like the aroma molecules that carry the scent of ingredients, similar molecules are responsible for the flavors of these ingredients. In fact, there is a large overlap between the two; several aroma chemicals are actually edible and responsible for the characteristic flavor of an ingredient.

As the world progressively embraces AI, one of the most intriguing applications is in the culinary industry. GenAi, extended to the domain of food, fundamentally transforms how recipes are created and food pairings are thought up. It can be used to devise a myriad of new culinary concoctions that are not only tasty but also in closer synchrony with consumer preferences.

Want a new flavor idea for chips, or sour candy? Use a GenAi powered engine that leverages consumer perception data and a diverse collection of preexisting recipes to produce perfectly tailored and unique flavor pairing ideas. To achieve such accurate precision in suggesting flavor combinations, such a model would rely on three distinct sources of information: a framework guiding it to extract meaningful and important flavors that would resonate with the consumer; a voluminous database of global recipes; and a comprehensive catalog of flavor molecules found in common food ingredients (Garg et al. 2018), FlavourDB: a database of flavor molecules (foodb.ca). Using this data, GenAi creates a unique flavor molecular network, where ingredients with common flavor molecules are strongly connected. These ingredients create the basis for establishing new, innovative recipes, with a high likelihood of favorable consumer response.

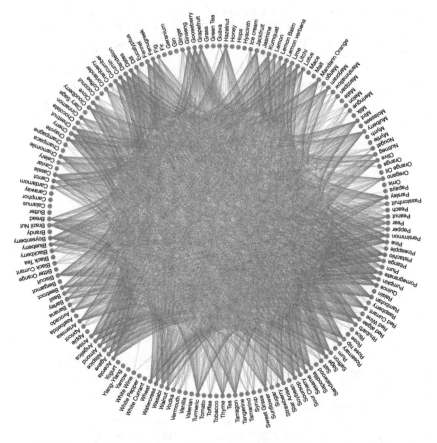

FIGURE 14.2   The figure plots a small subset of the flavor molecular network. Ingredients that share at least one flavor molecule in common are connected by a line.

An important principle that could guide such a system is the food pairing hypothesis, where ingredients that share a significant number of flavor molecules typically pair well together (Ahn Yong-Yeol et al. 2011). In the brain, neurons that fire together, wire together. By combining flavors that share these commonalities and by accounting for the popularity of ingredients in certain demographics, the paradigm of recipe creation is taken to an entirely new level. This unique approach ensures the development of truly appealing and innovative recipes, thereby redefining culinary creation with AI.

In this chapter we have presented a fundamentally different way to understand fragrance and flavor through associations in the brain. We have also presented a transparent path to designing fragrance and flavor that lets us not just rely blindly on expertise. This harmonious blend of neuroscience, algorithmic transparency, and combined human expertise will enable us to design captivating fragrance and flavors that are guaranteed to please consumers.

CHAPTER 15

musicGPT – Extract Emotions in Music and Generate Original Music

Music activates a whole series of neural pathways in the brain, especially those related to emotion and memory, among others, largely bypassing circuits involved with language.

That alone explains the fundamental power of music to evoke our emotions and summon happy memories of music that we have heard.

The soundtrack of our lives begins at an early age, and continues through life.

This chapter details how the brain processes music. Just as importantly, it discloses how marketers can – and should – use music as a strategic marketing mechanism. It outlines how GenAi can partner with neuroscience to make music applications even more evocative and memorable.

The Melodic Mind: Harmonizing Marketing and Neuroscience

Visualize the intricate and ever-changing network of neurons in the brain, comparable in complexity and dynamism to a symphony orchestra. Every section, including brass, strings, and percussion, contributes to the creation of something that exceeds the combined value of its individual components. In this intellectual symphony, music serves as a unifying element that intertwines with the cognitive and emotional aspects of the mind. Neuroscience has long acknowledged the distinctive ability of music to concurrently activate numerous brain locations, which becomes particularly evident in clinical studies of aphasic patients – individuals who experience difficulties with speech production and understanding. Interestingly, these individuals frequently maintain the capacity to sing (Rommerud 2016).

This observation underscores the deep roots of musical processing within the brain. Through avenues not yet fully understood, melodies bypass conventional language pathways, tapping into a primal circuitry of communication. Research indicates that these pathways cross through regions responsible for memory, motor control, emotion, and reward. When we rely on music as an adjuvant to reach the nonconscious, we activate an undercurrent of neural activity – silent to consciousness yet powerful.

Furthermore, cognitive neuroscience demonstrates that music is not merely a passive backdrop; rather, it imbues tasks with a qualitatively different mental state. While seemingly multitasking – working with music in the background – our brains reveal an impressive duality. One neural network maintains focus on primary tasks, while another seamlessly processes the melodies without conscious effort, illustrating the brain's capacity for parallel processing.

What does this mean for the marketer? The answer lies in leveraging the dual-processing nature of our brains – the conscious and nonconscious streams. During the first critical seconds of a commercial, for example, music can serve as a cognitive usher, guiding the nonconscious toward a receptive state. Studies have shown that these opening bars can anchor attention, set mood, and create lasting impressions that influence the overall effectiveness of an advertisement.

Be it lighthearted jingles or soul-stirring scores, auditory branding must align with the narrative at hand. Research affirms that music congruent with the brand's image amplifies consumer engagement, while discordant music can disorient and detach.

The seamless blending of music within an advertisement accentuates its narrative, helping the story unfold within the mindscape of the audience. Marketing application here is both a psychological puzzle and a creative challenge, striving to trigger the desired emotional and cognitive responses.

We must also consider the role of musical tempo and rhythm in shaping physiological and psychological reactions (Kim & Ogawa 2019). Rapid beats might increase heart rate and alertness – useful for exciting, high-energy products – while slower tempos lend themselves to relaxation and contemplation, befitting ads for calming or luxurious goods. Here, the alignment of music with the pacing of the narrative can enhance absorption and retention.

However, music's potency extends beyond the mere match-up of tempo and product. Its capacity to invoke nostalgia, to summon the ethos of a certain era, or to establish cultural relevance cannot be overstated. Through these mechanisms, brands can connect with their audience on a deeper level, anchoring themselves to memories and emotions that transcend the immediate messaging of the ad.

And yet, invest as we might in crafting the perfect melodic accompaniment for our marketing message, the power of music to bewitch and bedazzle should not lead us astray from our central narrative. While the notes can open the door to the nonconscious, they must not trample over the story we aim to tell. They are there to support, to enhance, not to usurp the leading role. Hence, the strategic planning of music in marketing should reflect a thoughtful composition – attuned to the flow and crescendos of the brand's tale.

Rhythm and Reason: The Neuroscience of Beats in Marketing

The intricate patterns of rhythm, essential elements of music, evoke a primal response in all humans. Like the steady beating of a heart or the syncopated rhythms of a dancer's feet, rhythms penetrate the layers of the conscious to stir something innate within us. Neuroscience breaks down this phenomenon by exploring how our brain perceives and processes rhythm – a sequence of timed beats – and its biological significance. One study reveals that rhythmic stimulation activates motor areas of the brain, leading to responses such as tapping one's foot or nodding one's head in time with the beat (Grahn & Brett 2007).

But why is rhythm so effective? At its core, rhythm speaks the language of the body. The synchronization of movement with rhythmic sounds, a phenomenon known as entrainment, illustrates how deeply interconnected our physical reactions are to auditory stimuli (Kotz et al. 2018). Consider a high-energy workout song: its driving beat not only motivates but also keeps pace, helping the exerciser align their movements with the pulse of the music. This synchronization results in not just a more effective workout but an inherently satisfying experience, as the harmony between body and beat creates a pleasurable feedback loop.

Brands can harness the power of rhythm by integrating it into the narrative of advertisements. When rhythmic elements sync up with visuals in a commercial, they create a more immersive experience for the viewer. The inherent structure provided by a solid beat can guide the storytelling, keeping the pace and ensuring the message is delivered in an impactful way.

Melodic rhythm, where melody and rhythm converge, further magnifies this impact. The appeal of melodic rhythm is exemplified by Beethoven's compositions, where structured rhythm couples with melody to evoke powerful emotions. In a marketing context, melodic rhythm can serve as a mnemonic device, making a brand's message more memorable; it's where the hook of the music becomes inseparable from the product being advertised.

It is vital, however, for marketers to ensure that the rhythm they choose aligns with the brand identity and message. A mismatch here can lead to confusion or disengagement. For example, a luxury car brand might opt for a smooth, understated beat rather than an aggressive, fast-paced rhythm. Similarly, a sports drink advertisement could capitalize on a high-tempo beat to amplify a feeling of energy and endurance.

The relationship between rhythm and consumer behavior extends to the realm of purchasing decisions as well. A study in retail environments showed that background music with a slower tempo led to increased time spent in-store, while faster music encouraged quicker shopping but also raised the likelihood of impulse purchases. It would seem that marketers have at their disposal, in rhythm, a metronomic tool with which to subtly influence buyer behavior (Zeeshan & Obaid 2013).

Rhythm doesn't only mean beats per minute, it encompasses the "how" of sharing a message. The "staccato" of billboards seen in passing, the "adagio" of a lingering online video, each has a rhythm of engagement. In the symphony of sales pitches, humor and wit might act as the playful pizzicato – the plucking of the strings – enlivening the dialogue and endearing the brand to its audience.

Furthermore, rhythm in marketing need not be an overt, audible beat. It can manifest in the pace of speech in a voiceover or the cadence with which a brand story unfolds across a campaign. The tempo of communication, the pauses between points, and the momentum built throughout can all contribute to a viewer's engagement and, crucially, recall.

Finally, the elegance of execution in how rhythm is used in marketing communications can reflect back on the brand itself. Marketing messages that are finely tuned – where rhythm, content, and delivery are in perfect harmony – can endear a brand to consumers in ways that transcend the immediate sales boost.

GenAi can analyze existing music tracks across different genres and extract rhythmic patterns. GenAi algorithms can also generate rhythmic patterns and beats based on input parameters such as tempo, genre, and mood. These rhythmic patterns serve as the foundation for creating engaging music tracks.

GenAi can analyze the emotional impact of different rhythmic patterns and recommend those that align with the desired emotional tone of the advertising campaign. By understanding how rhythm influences emotions, marketers can select rhythmic elements that evoke specific feelings in consumers.

Audio Visual Synchrony – Timing Is Everything

When sounds overlap, the brain must act as a conductor, orchestrating a balance between competing stimuli. Scientific research has illuminated the brain's remarkable ability to process auditory information, revealing critical findings about attentional focus. Neuroscientists have demonstrated that in the presence of both speech and music, the auditory cortex is met with a challenge: to process linguistic content and musical tones together. This simultaneous processing, however, can cause one to overpower the other, resulting in a phenomenon akin to an auditory tug-of-war.

The science of auditory masking explains how sensory inputs can interfere with one another, and in the world of advertising, this speaks volumes – literally and figuratively. When a voiceover competes with background music for dominance, the auditory system often tips the scale in favor of the music. This prioritization is rooted in our evolutionary history; the brain's affinity for melody, possibly a trait that helped early humans bond and communicate, may explain the magnetic pull of music on our neural pathways.

The implication for marketers is clear: to ensure that a message isn't just heard but also understood, the music's volume must be orchestrated in a way that it complements, rather than competes with, the spoken words. Neurologically speaking, achieving an ideal balance is key. When the background music is softer, the voice can become the soloist, allowing its message to be processed with greater clarity.

Moving from science to application, consider the delicate art of timing. Just as a well-timed cymbal crash punctuates a symphony, strategic silences within an advertisement can spotlight the spoken message. Studies show that the brain's attentional networks are highly sensitive to changes in auditory input, meaning the precise muting of music at key moments can amplify the impact of words.

Furthermore, the synchronicity between audio and visual elements in media is a dance that the brain follows intently. When the packaging does not match the product – when a voice doesn't align with the movement of the lips – we experience a cognitive dissonance that the brain strongly rejects. The consequence in advertising is a jarring experience for viewers, swiftly undermining credibility and engagement.

Cognitive psychologists have uncovered that the brain has specialized regions, such as the superior temporal sulcus, which are finely attuned to the temporal congruence of auditory and visual cues. Consequently, for marketers to craft compelling narratives that the brain readily accepts, the marriage of sound and image must be seamless.

The essence of communication in marketing – whether it's transmitting the cheer of a soda pop commercial or the sophistication of a luxury car ad – is captured not only in what is said, but how it is said. The rhythm of dialogue, the pause for emphasis, and the inflection of the voiceover artist contribute to a symphony of signals that the brain decodes. The marketer must direct these elements with precision to evoke the desired emotional resonance.

Sounds of Attention – What We Have Evolved to Pay Attention To

Life is enriched by its soundtrack – each note and rhythm carve memory and molds emotion. Science understands how auditory events pull the strings of our cerebral cores. Within our brain, the auditory cortex commands a powerful influence over our emotional responses to sound, as if it were a maestro cueing the rise and fall of our internal ensemble.

The concept of auditory ascendancy is compelling – sounds that ascend imply an increase in energy or a build-up of emotional intensity, tying neatly into the human instinct for upward movement as a positive progression.

Cognitively, the act of perceiving ascending melodies corresponds with increased activation in neural circuits associated with reward and motivation (Putkinen et al. 2021). This is where science transforms into storytelling – ascending pitches can spark aspiration in a clever advertisement. A car commercial that uses ascending tones as the vehicle accelerates, or a soaring scale as an airline jets off into the clouds, taps into this uplifting phenomenon with finesse.

Mirroring this ascent, sounds that descend require their own fanfare. The psychological effect of descending musical events echoes the visual cues of sinking or settling, often used to tranquilize the energetic tempo set by their ascending counterparts. These tones can represent comfort, resolution, or completion. Marketing applications are ripe for the picking; the soothing decrescendo of music as a luxury bed contours to the sleeper's body in a mattress advertisement or the calming drop in pitch to signify relaxation in a spa promotion.

Shifting to subtler sounds, whispers and murmurs exert a clandestine charm. Neurologically, whisper-like sounds trigger increased attention as we strain to extract meaning. This encodes a robust memory trace, as cognitive resources are enlisted for decoding. Moreover, evoking emotive sounds like moans and groans activates our brain's empathy circuits. The limbic system engages, interpreting these sounds as authentic emotional expressions. When presented in a strategic marketing scenario, such as sympathetic groans accompanying a frustrating situation that a product can alleviate, we're not just hearing a pitch; we're experiencing a scenario that tugs at our empathic strings.

Scattered throughout a commercial, laughter acts as a catalyst for social bonding. Laugh tracks in advertisements may sometimes appear trite, but they play upon deeply rooted neurological scripts that invite viewers to join in the mirth. As the brain processes these sounds, mirror neurons fire, and the innate social synchronization takes hold; laughter becomes a shared, unifying language.

The voices of children are potent symbols of purity and potential. Neuroscientific studies affirm that these sounds trigger protective instincts, due to which they deeply resonate with us.

In conjunction with these childhood melodies, rhythm is important in the realm of attention and memory. The syncopated beats in an advertisement don't just mark time; they craft a scaffold for emotional engagement and retention. Catchy jingles, with their stickiness, are threads in the tapestry of cultural zeitgeist. The catchy rhythm becomes the mnemonic that ensures the brand name hums loyally in the mind's echo chamber.

GenAi-driven music creation tools can generate compositions that express a wide range of emotions, including joy, excitement, sadness, nostalgia, or inspiration. Marketers can specify the emotional tone they want to convey, and GenAi can produce music that elicits those feelings in listeners.

GenAi can create melodies and harmonies that evoke specific emotions through their musical characteristics, such as pitch, rhythm, tempo, and dynamics. By manipulating these elements, marketers can craft music that resonates emotionally with the target audience.

Voice of Authority – The Role of Pitch, Tone, and Timbre

In the realm of auditory perception, the timbre of a voice carries more than a message; it unfolds a spectrum of cultural expectations and biological predispositions. Neuroscience unmasks our brain's nuanced response to gendered voices, revealing an intricate ballet of neural activity that dances to the different pitches and patterns bestowed upon us by nature. The auditory cortex, our central hub of sound processing, showcases a heightened sensitivity to these variations, suggesting a predisposed tuning to gender-specific vocal qualities (Weston et al. 2015).

The gender of a voiceover artist can sway the perception of the product being marketed. When we encounter a female voice in advertising a product for women, it may resonate more closely with the audiences' lived experiences. The genus of this connection lies in the way our auditory system is wired to prioritize voices that are biologically similar to our own, fostering a subconscious sense of empathy and understanding.

However, the interplay of gender and voice in marketing is nuanced. A male voice evoking the traditional attributes of authority and depth can introduce an intriguing contrast in a space typically reserved for female voices, captivating attention through the element of juxtaposition (McClean et al. 2018). Conversely, when the target audience is predominantly male, a female voice can weave a narrative that feels both nurturing and persuasive to the shopping experience.

In this auditory realm, the resonance of a voice often speaks louder than the words themselves. The perception of gravity and seriousness is often ascribed to lower-pitched voices. Such tonal gravitas has roots in evolutionary psychology, where depth of voice has been associated with strength and dominance – a factor that remains subtly ingrained in our subconscious processing (Zhang et al. 2019).

The portrayal of the "voice of authority" in media often deftly employs this principle. A voice that rumbles with the susurrus of grounded certainty commands our attention, coaxing us to associate the sound with importance and credibility. It's a symphonic stroke that marketers can utilize to underscore the significance of their message, delivering it with an auditory weight that resonates within the listener's cognitive auditorium.

Humor can pivot the perception of gendered voices. The surprising application of masculine depth to a lighthearted product or the sparkle of femininity in an unexpected context can jolt the listener out of apathy. The novelty tickles the brain's fancy for the new, making the message and, by extension, the brand, more memorable – a playful deviation from the expected harmonies of gendered communication.

Should this auditory alchemy skew too far toward repetition or monotony, however, neural pathways dim in disinterest. Cognitive psychology cautions against overusing any element of surprise. The essence lies in crafting a narrative arc that flows with subtle variations, maintaining the listener's curiosity while anchoring the story with familiar tones – an approach that balances novelty with the comforting caress of the known.

Transitioning from the scientific to the strategic, we learn that tonality and gender are instruments of human connectivity. To play them well, a marketer must not only know their audience but also feel the rhythm of their desires and expectations. It's a sonnet of persuasion, where the choice of narrator can strike a chord of trust or inspire a leap toward something bold and new. In the marketing symphony, the right voice at the right pitch can be the crescendo that lifts a brand's presence to new heights.

As marketers write the score for their next campaign, they might consider the rich vibrato of gendered voices as a palette from which to draw inspiration. Let the light-hearted tenor of a male voice selling children's toys weave its magic, or allow the assertive clarity of a female voice power an automotive campaign. These are not just strategies; they are the playful notes in marketing's grand composition, each one capable of stirring the hearts and minds of an audience waiting to be moved.

GenAi can be programmed to create music using synthesized voices that are perceived as masculine or feminine. Marketers should consider the appropriateness of each voice type for the target audience and the emotional tone of the campaign. Consider incorporating voices from diverse gender identities to reflect the full spectrum of human experiences. GenAi can be programmed to create music using voices that represent nonbinary or genderqueer identities, promoting diversity and inclusion in advertising campaigns.

Be mindful of cultural norms and sensitivities surrounding gender representation. Ensure that the use of gendered voices aligns with cultural expectations and does not perpetuate stereotypes or biases.

Extracting Emotions Evoked by Music – Anthropocentric Models

We attempt to answer a fundamental question that is thousands of years old – what are the precise emotions evoked by music in the brain? We posit that at its core, models that connect music to emotion are foundationally anthropocentric; the sounds we make when we are emotional, form the basis of how we associate sound and music to emotion.

Auditory stimuli evoke emotional landscapes within us. Research suggests that the auditory cortex is not just processing raw data but actively seeking emotional connection (Juslin & Laukka 2003). When we hear music that encapsulates the very essence of our vocalizations during emotional experiences, the brain generates a mirrored emotional response, creating a symphony of perceived feelings.

The amygdala dynamically contributes to the processing of music. It is tuned to the emotive qualities of the sounds we consider human-like, playing a critical role in generating our affective responses (Zhao et al. 2018). When a piece of music captures the timbre or pitch of the human emotional voice, the amygdala orchestrates an internal response that replicates the sentiment expressed in the music.

This phenomenon is underpinned by empirical evidence indicating that music can trigger the release of dopamine, similarly to other rewarding stimuli, hinting at an ancestral origin where sounds might have signaled socially pertinent events. Our neurochemistry responds to the rhythm of these sounds, subtly swaying our emotions and enhancing our mood in accordance with the acoustic mirror effect provided by the music.

Transitioning to the realm of marketing, a poignant narrative is often embroidered with a soundtrack that mimics the prosody of human emotion. Marketers can leverage this mirroring effect between music and emotional sounds, aligning a brand's message with the auditory echoes of the desired emotional landscape. The goal is to create a resonant experience that moves the consumer, harvesting the deep-seated connections between music and emotion to establish a memorable brand narrative.

Just as a catchy chorus can loop endlessly in our heads, a brand's jingle, when composed with an understanding of the brain's predilection for emotional sounds, becomes a recurring guest in the consumer's neural pathways (Jain & Jain 2016). The appropriate interplay of melody and emotion ensures that the echo of the brand's tune lingers, encouraging brand recall and affinity; this is the art of emotional branding in the realm of sound (Kellaris 2018).

Neural network models identify the emotional content in music of any length. We focus on emotion in music because it helps improve brand memorability and helps to more effectively communicate brand values and statements.

The development of our AI models was guided by a simple but powerful insight from neuroscience:

Our Brain associates a particular Emotion with a Sound when WE MAKE that Sound ourselves when expressing the Emotion

In keeping with this idea, our neural networks do not depend on human-generated tags for existing songs, but instead are trained using the fundamental nonverbal vocal utterances humans make when expressing particular emotions.

A set of human voices, along with several levels of musical *capture* and *extrapolation*, are used in the training of the models. Along with "joy," our models are trained to identify the following fundamental set of emotions that are felt by all humans and expressed in very similar ways across the world, as a set of core positive and negative emotions:

1. Anger
2. Disgust
3. Sadness
4. Pain
5. Fear
6. Irritation/Frustration
7. Courage
8. Desire
9. Joy
10. Love
11. Peace
12. Surprise
13. Humor

Thus, our models take any piece of audio and decompose it into a multidimensional emotional space across the length of the song. The audio file under consideration is first decomposed into a large set of audio features, usually numbering in the hundreds, through classical signal processing algorithms. These features then become inputs to the neural network models. The final output from the models is a second-by-second decomposition of the audio into its constituent emotions. Models can expand or contract on the core set of chosen emotions based on the necessary granularity. The order of the model can be chosen to represent with any level of compactness or expansion these global emotions.

In order to determine if an audio track aligns with the intended messaging or profile of a brand, we also need as input the descriptors of the brand profile. These may be words like "sophisticated," "bold," or "free." The emotional content of these brand words are determined using a semantic to mood space map, thereby translating the brand profile into the same multidimensional emotional space described above.

The mu score

The two parts of the system are represented in Figure 15.1. On the one hand, every second of the audio track is decomposed into the multidimensional emotions using the neural nets, while on the other the brand profile is mapped into the same multidimensional emotional space. This allows us to evaluate how close or far each second of the audio is from the target brand profile.

We are calling this measure of distance the *mu score* (mu for music). It takes values from 0 to 5, with 5 being most similar and 0 being the most dissimilar. Since this distance is evaluated across the entire length of the song, it allows a user to evaluate both the song on average, as well as identify sections of the song that resonate most with the brand profile.

Generating Music Evoking Particular Emotions – Music Transformers

The second piece of the puzzle is not simply to figure out if a piece of audio matches the desired intent but to actively generate pieces that can be quantitatively shown to be a good match. This is where the generative capabilities of AI systems come into play. Leveraging artificial intelligence models for music generation opens up an entirely new realm of possibilities. These AI systems are designed to learn and replicate the patterns it discovers in the vast datasets of music it's exposed to during training. In essence, it extrapolates numerous factors within music, such as melody, rhythm, or the emotive nuances in various songs to create original compositions. This enables the creation of unique melodies and harmonic structures that might not have been conceived by human composers.

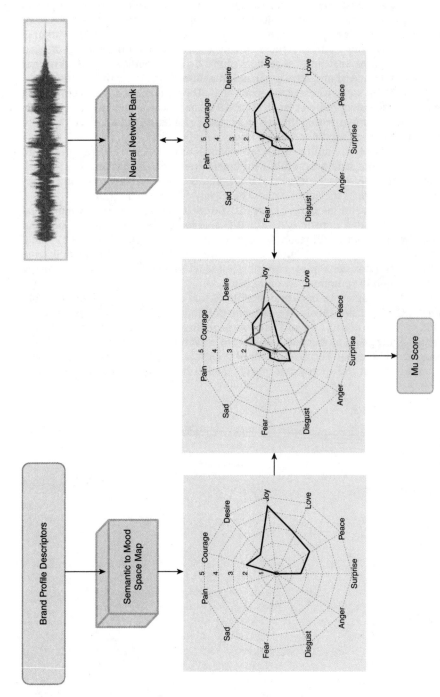

FIGURE 15.1 A schematic representation of how models can be used to map brand profiles to matching musical content.

AI-based music generation models take inspiration from a variety of genres and periods and then generate music by utilizing a complex set of algorithms that can understand, learn, and reproduce musical elements. The models commonly work by learning from sequences of musical notes or arrangements, or even audio spectrums, instead of sequences of text as in conventional LLMs. Music is inherently temporal and sequential, and therefore lends itself wonderfully to be "tokenized" in some way and fed to an LLM. These could be done by converting music to a sequence of notes (such as in MIDI), or by looking at the corresponding sheet music, or by treating the waveform itself as a "numeric sentence" (refer to the introductory chapter on LLMs).

These LLMs are built to analyze sequences or patterns in existing music and then predict and generate the following musical notes based on what it has learned. The end result is a machine-created piece of music that retains the essential structure and pattern of the genre but is distinctively original in its entirety.

Connecting the generative piece to the AI networks that determine emotional content provides a holistic system that is able to generate emotionally rich and complex music that will resonate as expected and pull at the correct emotional strings; this could be to design an effective sonic signature for a brand, or even to generate compositional pieces for movies and television that perfectly accentuate and heighten the emotionality of the scene.

In this chapter we have presented tools that can be utilized to capture audience attention using music. We have also presented models that can extract emotion evoked by music, and laid out how generative models can create music that matches the particular emotional cues of a brand or an ad. We also note that these techniques can be effectively used to either create copyright-free music or to choose the right and appropriate music to create the emotional backdrop for an ad.

CHAPTER 16

adGPT – neuroAi-Powered Advertising

Advertising is *aspirational*.
The best – that is, the most *effective* – advertising attracts us.
Involves us.
And is welcomed into memory by our nonconscious mind.
Done right, this connection with the consumer triggers *desire*.

The best advertising has always sprung from the singularly human capacity to create something new, appealing, and surprising. That divine spark has given us moments that we can recall long after they appear. This chapter is about how advertising is experiencing a whole new realm, where human creativity is augmented by breakthrough technology the likes of which the world has never seen before. The result is the potential to create advertising that more consistently – more *effectively* – captures our attention, engages us emotionally, and is granted precious entry into memory.

Neuroscience teaches just that; the key to truly effective messaging must have those three components: *attention, emotional engagement,* and *memory retention*.

This chapter reveals how the brain receives and processes advertising. It offers a unique entry into the startling combination of neuroscience with GenAI. It is this pairing that is taking advertising to new heights of creative effectiveness – art and AI interlaced together to produce messaging that the nonconscious mind finds irresistible.

neuroAi Nuance: Graceful Weaving of Story and Product

In the neural pathways of the brain, stories evoke a powerful response that has been etched into our cognitive processes over millennia. Narrative psychology suggests that humans inherently think in narratives, constructing our identities and understanding the world through a story-driven lens. The narrative form is compelling, as it helps organize complex information into digestible and emotionally impactful segments.

Transitioning from narrative to the hard lines of product specifications, neuroscience observes an emotional ebb. Neuroimaging technologies, such as fMRI, reveal that emotional engagement declines when a story gives way to a product-focused exposition. These changes in brain activity underscore the shift from holistic, narrative thinking to analytical, detail-oriented processing, orbiting away from the neural territories rich in personal connection. Rarely does effectiveness return from this precipitous dip.

It is therefore critical to switch the product to the background while the humanity and story is in the foreground, and have humanity and the story be in the background while the product is in the foreground. Thus, the abrupt discontinuity is gone.

Retention of narrative elements throughout a presentation or advertisement acts as a cognitive life raft for the audience. According to the continuity theory of narrative processing, maintaining story elements allows the audience to preserve mental models of events, sustaining engagement and comprehension even during technical expositions.

The act of juxtaposing human narratives with product features demands dexterity. Each element must be managed instinctively without causing cognitive or emotional overload. This balancing act, if mastered, ensures the seamless integration of features into the fabric of the story being told.

Humanizing products through anthropomorphism, assigning human characteristics to otherwise inanimate objects, can catalyze relatability and empathy. This scientific observation has profound implications in marketing and can act as a connector in the neural relay between character-driven and product-driven narrative shifts. Algorithms that create advertising must know how to weave the product into the story.

Subtle yet Present, Implicit Branding

From a neuroscientific perspective, our brains are wired to seek patterns and coherence in the sensory world. The concept of repetition blindness, a term introduced by Kanwisher (1997), presents an intriguing cognitive phenomenon where the repeated presentation of an item results in its reduced perception. This mental filtering mechanism suggests that consumers are adept at ignoring overt and repetitive brand placement. Thus, marketers must walk a fine line between recognizing the brand and overexposing it to the point of invisibility.

Branding in marketing is not just about visual representation; it is the integration of all senses into a unified brand experience. Sensory integration theory as postulated by Ayres (1972) suggests that the brain's ability to combine information from various senses is essential for understanding and navigating the environment effectively. For brands, this means the subtle use of colors, sounds, textures, and even scents could enhance brand recognition without the need to be overbearingly explicit.

The mere exposure effect, a psychological phenomenon described by Zajonc (1968), suggests that familiarity leads to preference. This effect highlights the role of indirect and ambient exposure of brand cues in fostering brand familiarity. Exposure to intrinsic cues related to a brand – such as colors, shapes, or sounds – spread throughout an advertisement, could subconsciously scaffold recognition and recall without overwhelming the consumer's conscious attention.

The key then is evoking brand presence without presence. Implicit memory's role in human cognition suggests that learned experiences can influence thoughts and behaviors, even when one is not consciously aware of the memories. Concepts like brand presence can benefit from this cognitive underpinning by becoming implicit through the surrounding narrative – hiding in plain sight, so to speak, fostering brand awareness through association rather than direct exposure.

Transitioning from theory to practice, we understand that subtlety is paramount. Rather than plastering a logo across every frame, the use of brand colors or themes in background elements – a coffee mug, a street sign, or a character's clothing – can gently engrave the brand on the viewer's mind. Think of it as an elegant ballet of brand elements, each performing in the narrative space without stepping on the toes of the storyline.

Just as a skilled composer creates music that sticks with you long after it's played, a marketer composes an ad with the brand as its recurring leitmotif. Subtle references to the brand rhythmically strewn through the ad ensure that the narrative is both harmonious and memorable, fortifying the brand's story with each beat.

In today's digital age, where consumers are adept at skipping ads or blocking them outright, subtlety in branding becomes even more crucial. By integrating the brand into shareable, relatable moments in an ad, marketers can turn passive viewers into active promoters. The brand becomes part of the social conversation, not as a shout but as a whisper that beckons attention.

In the implicit-branding approach, the soft touch of branding elements, like the faint echo of a distant melody, imprints on the viewer without overtaking the scene. Contrary to explicit branding methods, implicit cues are more likely to bypass the individual's tendency to resist overt persuasion, known as reactance. This method leverages the power of subtle cues that the reticular activating system filters as less threatening or sales-driven, thereby fostering a positive brand outlook.

There come moments in advertising when the product must be brought to the forefront. Here, too, the brain prefers a subtle human touch, even if only implied, as evidenced by research on neural responses to object-human associations. In instances where the product takes center stage, behind-the-scenes humanity must sustain the emotional environment, ensuring continuity in the consumer's experience.

Marketers can leverage this understanding, embedding brand hues in a protagonist's attire or the background score's notes, to strategically place the product within the consumer's emotional theater.

Composing and Activating the Sensory Iconic Brand or Product Signature

The human brain is remarkably adept at creating associations between sensory experiences and emotions, developing shortcuts that evoke entire narratives with minimal cues. This cognitive feat rests on the astonishing capability of our brains to form and retrieve these associations from sensory inputs. The fizz of a soda, the silhouette of a bottle, the smell of freshly brewed coffee – all these can anchor a vast web of associations that convey the essence of a brand. The activation of iconic signatures allows connections to the brand to flourish.

An iconic signature serves as a mnemonic, a powerful tool for enhancing both the memorability and recall of brand experiences. Through repeated exposure, a sensory signature becomes imprinted in the consumer's memory, connected not just to the product but the emotional landscape it inhabits. Be it the pop of a beverage can or the rich aroma wafting from a hot cup of coffee, these sensory signatures can bypass the need for elaborative cognitive processes, triggering immediate recognition and affective responses.

Ritual in consumer experience is rooted in the habitual patterns and routines encoded within our neural architecture. The daily rituals we engage in – like sipping morning coffee or enjoying an evening beer – can create and reinforce pathways in the brain that endow these acts with a sense of comfort and normalcy. Thus, identifying and incorporating the ritual surrounding a product deepens the connection between the consumer and the brand.

To activate an iconic signature is to engage the multisensory experience in a marketing context. The neuroscientific principle behind this is cross modal correspondences – where one sense can influence or even alter the perception of another (Teichert & Bolz 2018). For instance, the sound and sight of a can opening can heighten the expected pleasure of taste. When such sensory signatures are activated within an advertisement, they not only capture attention but can also intensify the craving for that multisensory experience.

Simplicity is an allure that the brain finds hard to resist, and sensory signatures are elegantly simple cues. Our brains are innately drawn to the path of least resistance, a principle known as cognitive economy. An iconic sensory signature simplifies the plethora of product information into a single, easily digestible experience. These sensory shortcuts not only streamline cognition but also craft a brand's allure that is as effortless as it is robust.

Clink, Sip, Ahh: Embodying the Ritual. Once the sensory hallmarks of a brand are distilled, infusing these into advertisements can lead to what's termed embodied cognition (Yim et al. 2021). This phenomenon allows consumers to vicariously experience the sensory gratification of a product – like the clinking of glasses or the first sip of a warm beverage – through mere observation. Brands aim to elicit embodied responses through careful synchrony of sensory cues that align with consumers' personal rituals, thus amplifying the allure of the product experience.

Activation of sensory signatures and rituals in advertising should not be cacophonic but rather a symphony calibrated to stir the consumer's sense and sensibility. It's about finding the natural rhythm of the product experience and amplifying it through clever and tasteful sensory integrations within the ad.

Changeup the Beginning of an Ad – Stress versus Opportunity

Most advertising has the following paradigm: problem, complication, resolution, brand. The early moments of the ad dwell on the problem to which the brand becomes a solution. The early moments of most ads, in amplifying the "problem," fill the viewer with empathized and perceived stress.

The human experience is profoundly shaped by the brain's response to stress, as much as it is by the attraction to opportunity. Neuroscience has shown that stress triggers the hypothalamic-pituitary-adrenal (HPA) axis, culminating in the release of cortisol – a hormone that, while useful for survival responses, inhibits cognitive functions related to goal-directed behavior (Kluen et al. 2017). When stress is perceived, the brain prioritizes the fight-or-flight response, which can overshadow the capacity to perceive and pursue opportunities.

While stress responses are crucial, humans are equally wired for opportunity-seeking. The neurotransmitter dopamine is vital in this regard, playing a significant role in reward and motivation circuits in the brain. When an opportunity is framed positively, it can stimulate dopamine release, leading to increased attention, enhanced creativity, and a tendency to engage rather than withdraw – an antithesis to the stress-induced freeze, flight, or fight response.

In advertising, the traditional approach of highlighting problems can inadvertently activate the brain's stress response. Yet, there exists an alternative. Neuroscience suggests that the brain is attracted to narratives that feature development and achievement (Sanders & Van Krieken 2018). By framing an advertisement as a journey toward an enlightening opportunity rather than a stress-induced struggle, it shifts the brain's focus from threat avoidance to goal pursuit.

There is a caveat to the avoidance of negatives in marketing: the undeniable memorability of negative experiences. This is attributed to the brain's amygdala, which significantly influences the consolidation of memory, particularly emotional events. While fear-based ads can be memorable due to amygdala activation, repeated exposure can lead to desensitization or aversive conditioning, thus leading viewers to disengage over time.

According to self-efficacy theory, individuals are more likely to engage in behaviors if they believe they can perform them and that they will lead to desirable outcomes (Schwarzer et al. 2008). By highlighting opportunities for empowerment, ads can harness the innate desire for self-fulfillment and personal growth, positioning the brand as a catalyst in the consumer's heroic journey.

Modern neuromarketing has embraced the transition from problematic to opportunistic narratives. Marketers now recognize that engaging with the aspirational aspects of human desire sparks a much deeper connection than merely offering solutions to stress. This opportunistic odyssey allows brands to bond with consumers on a journey of discovery and triumph.

As we ride the neural tide from empirical understanding to practical application, let's imagine ourselves as surfers on an ocean of brand opportunities. The latest trend in advertising shows a playful lightness – the liftoff over the wave rather than the crushing impact of the break.

Loved That Ad – But Wait, What Was It For? I Don't Remember

The effectiveness of a well-woven narrative is deeply rooted in neuroscience. As the brain encounters a story, cognitive coherence arises from the structured flow of events, enabling easy assimilation of the content. If a product is to become an authentic element of that coherence, its placement must resonate with the existing mental schema. A story that accommodates a product as a natural facet leverages the brain's predisposition for coherent, structured information, thus promoting integration rather than disruption.

Embedding a product within a narrative invokes the science of source monitoring, where the origins of memories are encoded alongside the content. When a product is congruent with the storyline, it becomes intricately tied to the narrative's origin in memory, creating a strong associative bond. These memories are more accessible, allowing the product to command a prominent place in the consumer's mind.

The intertwined nature of the human cognitive system and physical experiences, known as embodied cognition, explicates the seamless integration of products within a story. As narratives often evoke sensorimotor responses, a product that is interwoven, enhancing the tangible aspects of the story, becomes part of the lived experience, resulting in a deeper, more holistic consumer engagement.

The principle of saliency is integral for a product to become an organic part of the storytelling experience. When a product is salient, it interacts with the attentional networks in the brain, magnifying its significance. By aligning a product's features with key story elements, we enhance its neural salience, sensitizing the consumers' brains to its presence and importance.

As the narrative unveils, opportunities for product integration should be deeply embedded in the product storyline. When this is done with finesse, the product transforms from a mere cognitive concept into an activation ripe for consumption. The dopamine system, modulating reward processing, gears up when a cognitive representation aligns with consummatory behavior. The integrated product is then seen not as an intruder but as an indispensable ally in the story.

Let us now unpack the methods of product "unboxing" within the story. Instead of a gratuitous display, the product serves a purpose within the narrative, just as our brain luxuriates in finding purpose in actions. By contextualizing the product within the story's arc, its presence feels intuitive rather than invasive, guiding the consumer to an unspoken yet understood value.

Ads That Say Less: Fewer People and Places Say More, Say It Louder, and Say It Better

Our social brain revels in the tapestry of human connection, seeking to understand the subtleties of emotional exchange. It is in the depths rather than the breadth of these relationships that the brain finds gratification, carving out neural pathways for empathy and prediction. This neurological fondness for familiarity encourages us to return to circles where emotional expectations are met with consistency and familiarity.

Human cognitive capacity, according to Cowan's embedded-processes model, typically limits us to maintaining only a handful of informational chunks in our working memory. In the context of forming emotional connections, this translates to a preference for fewer, well-developed characters over an overwhelming ensemble. Exceeding this cognitive budget can lead to mental juggling, where no single character holds the spotlight of connectivity or memorability for long.

Advertising stories must involve no more than three characters – no big family gatherings and the family dog. The brain seeks to understand, connect and relate to a person; it cannot relate to a mob. So the fewer the number of characters, the better it is.

Do not transition from the living room to the playground, then to the office and to the dining room. The brain prefers fewer places, not an infinity of terrains and geographies to master, and comprehend in a few seconds. Keep it simple, and you will save your budget as well – without having to go all the way to New Zealand to film a simple scene.

Just when the characters in an ad have spoken, do not introduce a disembodied voiceover that talks instead of the voices we have heard. Remember to keep it simple – no cognitive overload.

Emotional investment in a character, akin to a financial stake in the market, flourishes with focused attention. Neuroscience corroborates that empathically connecting with fewer characters enhances quality and depth. The neural expenditures on empathy yield greater introspective understanding, leading to a stronger emotional payoff than the thin spread of attention across many characters.

Pleasure in social interaction often springs from predictability. Knowing how a character is likely to react provides a psychological safety net (Gerten et al. 2022). This neurological comfort zone is where audiences feel at home, enjoying the simplicity of social scripts that don't require constant cognitive recalibration. The richness of a few well-known characters can surpass the fleeting fancy presented by a crowd of extras.

In storytelling as in ads, crafting a narrative nest with a limited set of characters allows audiences to settle in, wrap their minds around the emotional fabric, and nestle into the familiarity. The cognitive ease with which we can snuggle up to the emotions of three characters over 30 means a more visceral connection, a longer-lasting impression, and often, a greater desire to return for more.

In advertising, the sweet spot is the intriguing balance where less becomes profoundly more. A commercial with a cast larger than our neural stage can accommodate becomes mental white noise. Adhering to a trio maximizes emotional saturation and audience retention, avoiding the cognitive chaos of a character caper. In this arena, three's company, and more's a crowd.

As we transition into the light-hearted realms, it's important to remember: In the world of advertising characters, think of adding people like adding salt to a dish – a little enhances, but too much makes the food inedible. Minimalism has its benefit; it allows for the audience's affection to indulge in fewer, richer characters, yielding a better recall. Here, it pays to be minimalistic in the construction of the narrative.

The friendly face, the reliable confidant, or the quirky girl next door are your emotional anchors. As the advertisement concludes and the curtain falls, these characters, now familiar friends, have carved their stories into the audience's psyche, warming their way to a home in a place we all want to go.

How Deep Tones Convey Wisdom: Finding Meaning in the Timbre of Voiceovers

The auditory allure of deep tones dates back to evolutionary biology, where lower frequencies were often associated with larger, more dominant animals and thus commanded more attention. The human brain, highly attuned to nuances in sound, instinctually associates deeper vocal tones with authority and wisdom. This psychological underpinning explains why traditional figures of gravitas, from tribal leaders to cinematic deities, are often endowed with robust, resonant vocals.

Sound pitch extends deeper into the psychological realm, engaging the subconscious with an efficacy that transcends mere pitch. Studies in psychoacoustics demonstrate that deeper tones activate the temporal lobes and the auditory cortex, where processing of emotional content in sound takes place (Moore 2014). This could explain why lower voice pitches are often instinctually perceived as having more significance, subtly swaying our emotive perception before even a single word is understood.

The pervading "voice of God" approach in voiceovers isn't merely a fancy; it roots itself in anchoring theory, where the first piece of information – like a resonant voice – serves as an anchor for everything that follows. An ad employing this technique could be tapping into deep-seated archetypes of wisdom and omnipresence, suggesting the notion that the voice, and by extension the message it conveys, is of substantial importance and reliability.

Beyond pitch, the artistry of voice lies in its timbre, tone, and tenor. The distinct tonal quality, or the timbre, of a voice is a rich tapestry that the brain processes as an acoustic fingerprint unique to each individual. When a brand selects a voiceover, the timbre plays a crucial role because it aligns with the brand identity as much as aesthetics do in visual branding. A warm and rich timbre can be as distinct and memorable as a logo.

When integrating voiceovers with on-screen content, a harmonious synergy is key. Cohesion between the voice and the visible character can anchor audience understanding in the narrative. A misalignment can generate cognitive dissonance, detracting from the message's potency. It's much like listening to a symphony where one instrument is out of sync; the entire composition's impact is compromised.

Leavening the profundity of deep voiceovers with humor provides a delightful juxtaposition. Laughter resonates on a biological level, releasing endorphins and serotonin, creating a pleasant association with the auditory experience.

A voiceover that resonates can personify a brand, lending character and depth that extends well beyond the close of a commercial. Crafting this auditory signature requires an understanding of the neuroscientific principles discussed while staying attuned to the humor and nuances of human preference. Employing a finely tuned, gender-aligned, robust voice, or the strategic absence of one, can cradle the audience in a narrative that feels both profound and personal, echoing far into the chambers of memory and consumer loyalty.

No Surprise, Women Are Tired of "Mansplaining"

The phenomenon extends into the sphere of gender identity, where the gender of a voiceover can impact how messages are received and internalized. Cognitive resonance with gender-matched voiceovers springs from same-sex modeling principles wherein individuals prefer and learn more efficiently from people they perceive to be like themselves (Casado-Aranda et al. 2018). This neurological alignment with our internal gender identity not only shapes our preferences but can also enhance the trustworthiness and relatability of the voiceover narrative.

The implications of gendered voiceovers in marketing are robust. Through empathy and identification mechanisms, female consumers tend to respond with heightened engagement to female voices. This improved relatability and identification could amplify the reception of the message, making it crucial to align the gender of voiceovers with the intended audience demographic, particularly in narratives where consumers are meant to see themselves reflected.

The Decelerated Delight – Slow Motion and the Mesmerized Mind

The human brain is inherently a predictive machine, constantly forecasting future events based on past and present stimuli for survival and efficiency (LeDoux & Daw 2018). When we witness events unfolding in slow motion, the temporal dilation enhances our predictive success rate, triggering a cascade of rewarding biochemical responses, including the release of dopamine, a neurochemical associated with pleasure and reward.

Our cerebral cadence is captivated by the deliberate pace of slow motion. The suprachiasmatic nucleus, often referred to as the brain's "master clock," regulates circadian rhythms and perceives the passing of time (Patton & Hastings 2018). Slow-motion imagery conspires with this neural timekeeper by elongating the moment-to-moment experiences, sharpening our anticipatory skills, and teasing out the joy of temporal prediction.

Anticipation, when paired with precision, forms the cornerstone of the prophetic appeal of slow motion. This visual phenomenon magnifies our neurological delight in accurate event prediction. As everything is decelerated, the brain bathes in the glory of its own prescience. These small victories in forecasting are not simply satisfying; they affirm our cognitive abilities in a world constantly in flux.

Slow motion acts as an attention anchor, exerting a hypnotic hold over its viewers. The prefrontal cortex, associated with focus and attention, is markedly sensitive to changes in visual temporal patterns. When the typical rhythm of events is slowed, this attentive machinery clings to the languid progression, encouraging a state of heightened sensory awareness, which not only captures but also retains our gaze.

Slow motion finds potent application in the world of culinary advertising, with the pour of a drink transformed into an elixir in deceleration. The sight of a liquid's languorous locomotion is not just aesthetically pleasing; it resonates with the brain's mirror neurons, which fire when we observe an action and imagine ourselves performing it.

The Neuroscience of Contact: Touch, Hug, and Hold to Connect

Serotonin, a neurochemical associated with feelings of well-being and happiness, surges with acts of connection. This *serotonin embrace* catalyzes a sensation of comfort and fulfillment – the neural correlates of what we feel during moments of intimacy with parents, children, pets, and even inanimate yet cherished possessions.

The appeal of puppies and babies in advertising goes beyond their innate cuteness. The human brain is hardwired to respond with nurturing instincts toward characteristics exhibited by infants and juvenile animals – large eyes, round faces, and soft contours. It's no coincidence that these "huggable moments" induce a cascade of serotonin release, creating a bond between the viewer and the visuals, rooting the advertisement deep within the limbic system.

Empathy's neural substrate is mirrored in advertising. The mirror neuron system reacts not only to actions we observe but also to the tactile experiences we see others undergo. When an advertisement features humans in a loving embrace or a hand tenderly resting on a pet, the viewer's neurons fire as if they themselves were partaking in the exchange. This instant neural mirroring is a cornerstone in forming the empathic connection advertisers seek to exploit.

The notion of humanizing products in advertising takes emotional transference to a whole new level. When a consumer witnesses a product being cradled or caressed, there's a transmutation of neural activity that imbues the product with life and warmth. This kinesthetic empathy shifts the consumer's perception of the product from the kingdom of the inanimate into the realm of the almost-human, generating caring and protective associations akin to those we experience with animate beings.

There's a neuropsychological basis for such anthropomorphism. The human fusiform gyrus, particularly attuned to facial recognition, is easily coaxed into seeing faces in nonfacial objects, such as the headlamps and grills of cars (Kühn et al. 2014). By assigning these quasi-emotive attributes to vehicles, ads subliminally compound the huggable dimension with powerful character narratives, whether lovable or aggressive.

The mere sight of touch can stimulate a tactile illusion in the observer, sparking a yearning to experience the sensation oneself. Research in multisensory integration suggests that visual cues can evoke somatosensory responses. These illusory phenomena are exploited in advertising to kindle the desire for contact. Showcasing a person gently stroking a plush blanket or the palpable quality of a leather seat in a car can virtually evoke sensory experiences in consumers, driving them closer to purchase.

Comfort lies at the heart of contact, and advertising narratives capitalize on this fundamental human need. By integrating scenes of touch – whether between people or with products – ads create narratives that echo life's most comforting moments. The warmth of parental embrace, the exuberance of cross-species friendship, or the assurance of hands-on craftsmanship – each molds an emotive arc that entwines the observer's personal experiences with the brand story.

People holding hands or hugging each other stimulates powerful emotions. The gentle holding of the face of a child by a parent or the affectionate touch of two people in love builds powerful emotions. These release serotonin and oxytocin in the brain from just observation alone. Another powerful neurotransmitter release happens when humans and pets hug each other. Be it as a static image, or a video, these visuals are powerfully emotive and evocative.

The gentle exploration and sensual experience of the product – we call it "romancing the product experience" – is yet another powerful attractor for the brain.

These principles, when used by GenAi, can produce powerful advertising content that aligns with what the brain seeks and derives pleasure from.

First Impressions Are All That Matter – First Five Seconds of Advertising

Research reveals that early engagement within the first seconds of stimulus exposure can determine attention sustainability. The five-second gateway not only serves as an opportunity for attention capture but sets up a cognitive anchoring framework that influences the sustained neural commitment to the advertisement narrative.

Sensory elements, when harmonized perfectly, have the power to orchestrate a gripping sensory attention within the primacy of an advertisement's debut. Studies in multimodal attention reveal that early auditory cues can significantly augment visual attention if paced rhythmically within the first five-second window. This auditory grip serves as a beacon, guiding the neural circuits of the brain to lock in on the audiovisual feast that unfolds.

Rhythmic entrainment describes how neural oscillations synchronize with external rhythms, as seen with auditory stimuli. Rhythmic cues in the first five seconds of an advertisement can create a "cognitive hook" by cuing neural oscillators to fall into sync, paving the way for heightened concentration and an emotional carousel that is hard to disengage from.

Rapid emotional engagement is the heart of the five-second lore in advertising. The amygdala exhibits rapid activation in response to emotionally potent stimuli. Introducing an emotionally resonant musical score or auditory cue in the early throes of an ad can trigger this amygdalar responsiveness, aligning consumers' emotions with the brand narrative from the outset.

Although the confines of a five-second window may seem restrictive for storytelling, the unfolding narrative can be effectively scripted to maximize this brief interlude. Efficient utilization of this sliver of time aligns with the brain's preference for chunking smaller units of information for swift processing and easy retrieval. A well-crafted start primes the neural pathways for the full story, positioning the introduction as a standalone micro-narrative with a macro impact.

Resonating music that emphasizes beats in the initial seconds of an advertisement not only captures attention but also pivots toward longer-term memory formation. By harnessing the auditory power of rhythm and beat, ads stimulate the hippocampus, a neural region implicated in memory consolidation. Emotionally charged rhythmic stimuli during this period can greatly enhance the likelihood of the ad – and therefore the product – being remembered.

High-impact advertising doesn't inherently demand high-budget production. Even the most fiscally conservative ad campaigns can employ strategic rhythm and music to ensnare the brain's auditory and emotional faculties within the first five seconds. This alchemy of rhythm presents a cost-effective tool that, when cleverly composed, can deliver the clout of their costly counterparts sans the grandiose visuals.

The Awe of the Unveil: Neuroscience Behind the Curtain Rise

The dance of anticipation begins even before the curtain rises. The human brain relishes the thrill of the pending reveal, activating structures associated with anticipation, such as the nucleus accumbens (Galtress & Kirkpatrick 2010). This anticipation is a cognitive tease, playing a neurochemical game of peek-aboo, as dopamine levels surge, heightening our state of expectant arousal. When the product – or any surprise – is finally unveiled, this orchestrated interplay of waiting and reward gratifies our neural circuitry.

Curiosity, as a cognitive emotion, is integral to the human experience; it compels exploration and learning. The precuneus, a region involved in episodic memory and visuospatial imagery, lights up in the face of novelty or partially hidden objects. This neurological response underlies the age-old fascination with the act of uncovering or revealing. A carefully choreographed product reveals hooks into our innate desire to discover, enhancing engagement and memory of the experience.

The visual cortex, which processes visual information, is tantalized by that which is not fully revealed. The "less is more" concept in neuroscience speaks to the way our brains fill in gaps and resolve incomplete stimuli. This partial information creates a suspense that is virtually irresistible. When a product is slowly unveiled, this sensorial flirtation galvanizes the visual cortex, leading to stronger neural imprints and making the subsequent full reveal vastly more compelling.

Expectancy violation theory posits that our brains are wired to detect and respond to things that deviate from the norm. The slow, dramatic revelation of a product defies our standard rapid pace of information consumption. This deviation from the expected rate of stimulus exposure captivates our attention and intensifies the memorability of the event. The deliberate pacing of the "curtain rise" engenders a disruptive momentum that the brain can't help but follow with intrigue.

In the art of advertising through the grand unveiling, creators juxtapose the neuroscientific underpinnings of wonder with the whimsy of anticipation. This delicate balance guides the consumer's journey from curious spectator to enchanted participant. Through the playful yet precise orchestration of the reveal, the everyday act of showcasing a product transcends into theater, cementing its impression in the consumer's mind.

Tackling Territorial Transitions: Navigating the Mind's Map in 30 Seconds

The hippocampus, associated with spatial navigation and memory, enables us to mentally map and navigate through different terrains. Rapid shifts in setting within a 30-second advertisement can overtax this neural navigation, leading to what is termed *cognitive dissonance* – a mental discord resulting from conflicting or disjointed information. In a brief timespan, the mind struggles to form coherent maps, which impacts the ad's overall effectiveness.

The art of scene harmony in advertising relies on the brain's capacity for scene integration. Research in perceptual psychology suggests that the brain prefers continuity and can easily become confused by abrupt transitions. When creating short ads, designers must choreograph scenes such that the images transition fluidly to prevent cognitive dissonance. This synchronization allows the brain's processing of landscapes to remain connected, preserving viewer's engagement.

Just as vital as geographic continuity is the careful handling of different time frames within an advertisement. The brain structures time into a coherent narrative flow, a process managed primarily by the prefrontal cortex (PFC). To avoid cognitive overload, advertisements should limit shifts in timelines, making transitions between past, present, and future exceedingly clear. Simplifying time-scapes within the ad's brief span prevents neural confusion and maintains the audience's story tracking.

Spatial coherence refers to the brain's preference for a consistent storyline within a unified space (Ushiro et al. 2022). Neuroimaging studies show increased activity in the medial temporal lobes when subjects perceive consistent story spaces. Advertisements can use this knowledge to create backgrounds and settings that change subtly, if at all, preserving narrative consistency and allowing viewers to maintain their spatial bearings amidst the storyline. The goal is to craft a story space that embraces the mind's geographical comfort zone.

Timeline transitions must be managed with chronological clarity to prevent cognitive dissonance. The lateral intraparietal area (LIP) of the brain is implicated in the perception of time and its orderly progression (Davranche et al. 2011). Ads that jumble time frames within their brief duration risk disorienting this cognitive clockwork, eroding the narrative's grasp on the viewer. For maximum clarity, temporal cues should be unambiguous, fostering a straightforward chronology that is easy for the brain to follow.

Combining knowledge of times and territories, advertisers can formulate neuroscience-based strategies. The goal is to align ad design with the brain's inherent narrative structure and spatial logic. To transform dissonance into harmony, advertising professionals employ practical techniques that align story elements. Visual storytelling, such as a character journeying from scene to scene, can bridge disparate geographies. Simplifying storylines that cater to the brain's chunking ability can create coherence and reduce the risk of cognitive dissonance.

A well-choreographed advertisement lightens the cognitive load, much like a graceful dance elevates the spirit. Techniques include signaling transitions with consistent visual themes or using a memorable jingle to introduce new scenes, catering to the brain's ability to process and enjoy patterns. This approach elevates the ad from mere information to an enjoyable experience, enhancing audience retention and appreciation.

GenAi that utilizes the principles outlined in this chapter will produce memorable advertising. Note that these rules must be embedded not only in story creation, but in the production of the ad itself. Be it GenAi or an agency that produces advertising, using these principles grounded in neuroscience will ensure that the ad will test well in copy testing, and will elevate purchase intent and brand recall; it will hold viewers' attention in a busy commercial break. As advertising transitions to digital media, the methods outlined in this chapter will make the ad stand out in the clutter.

CHAPTER 17

scriptGPT – Writing Entire TV Shows, Plots, and Dialogue

W hat sells a script?

How does a pilot become a network success?

What takes a genre film to box office stardom?

The key, of course, is audience appeal.

That mysterious, unpredictable alchemy of words and music and characters and plotlines and more that can produce an often unexpected hit.

In this chapter you will gain the knowledge and the tools to create filmed entertainment that strongly attracts and retains viewers as well as the neuroscience behind the process.

You will also learn how GenAi can supercharge the creative process – as a partner, not a replacement.

The Persona Palette: Neuronal Narratives in Archetypal Depictions

The exploration of character archetypes in storytelling is no mere writer's concoction; it is rooted in psychology and the neural processes of the human brain (Jung 1959). Swiss psychiatrist Carl Jung first theorized the concept of archetypes, which he believed were universal, archaic patterns and images that derived from the collective unconscious and manifested in individuals across cultures. These character molds have been integral in narratives throughout history, from ancient mythologies to modern cinema. The recognition of these archetypes taps into deeply ingrained neural circuits within our brains, eliciting an instinctive response from the audience.

Neuroimaging studies reveal that when we observe a character that fits neatly into an established archetype, certain brain areas associated with social cognition and empathy, such as the medial prefrontal cortex (mPFC) and temporoparietal junction (TPJ), are activated (Bzdok et al. 2013). These regions are involved in understanding others' intentions and emotions, enabling us to predict behavior based on our societal constructs of different archetypes. This speaks to why archetypal characters often feel familiar to us and why they resonate on a visceral level.

Of the numerous archetypes, Carl Jung's characterization of the hero, the rebel, the explorer, the ruler, and more, exemplify how these staples of storytelling also mirror neuronal pathways and expectations of behavior. For instance, engagement with a hero character often instills a sense of motivation and journey, while the nurturer may trigger emotional centers that evoke sympathy and protection. The villain, on the other hand, tends to activate areas related to processing threat and fear, such as the amygdala.

The concept of archetypes provides a map for writers to design multifaceted yet individually consistent characters that audiences can identify with, allowing for a structured approach to storytelling. When writers create characters that adhere to specific archetypes, they are in fact weaving a neurologically cozy narrative fabric, one that resonates with the audience's inherent processing of social and emotional cues (Anderson 2023).

On the flip side, characters that deviate from their traditional archetypal traits can trigger a jarring response in the viewer, generating a cognitive dissonance as the brain's predictability and pattern recognition systems (located in the basal ganglia and hippocampus) are challenged. These moments can be pivotal in gripping the audience's attention and generating discussions as viewers reconcile their expectations with on-screen actions.

This understanding has potent applications for screenwriters and marketers alike. By assigning archetypes and tracking portrayal consistency, media can capture the rapt attention of their audiences or purposefully subvert expectations to keep them on their toes. Analyzing audience response through measures like television ratings can unearth the neural predilections of viewers, illuminating the preferred alignment of characters to their archetypal nature.

The interplay between a character's archetype and their actions can be a compelling storytelling approach, influencing the emotional and physiological response of the audience. For instance, when a nurturing character unexpectedly betrays trust, this stark deviation can foster a memorable plot point, embedding the storyline in the viewer's memory through an intense emotional and cognitive experience, often leading to a spike in audience engagement metrics.

As an example, here is a specific set of archetypes that can be traced across movies and TV shows from all countries. Of course, characters can be complex and might embody elements across a couple of archetypes.

Types of standard male heroic archetypes:

Leader – a dynamic leader, he has time for nothing but work.

Bad and dangerous – dangerous to know, he walks on the wild side.

Loyal friend – sweet and safe, he never lets anyone down.

Lost and soulful – a tormented being, he lives in solitude.

Seducer – a smooth talker, he creates fantasies.

Scholar – coolly analytical, he knows every answer.

Showman – Mr. Excitement, he's an adventurer.

Warrior – a noble champion, he acts with honor.

Types of female heroic characters:

Boss – a real go-getter, she climbs the ladder of success.

Seducer – an enchantress, she charms to get her way.

Spunky and wild – gutsy and true, she is loyal to the end.

Free and dreamy – an eternal optimist, she dances to unheard tunes.

Geek – controlled and clever, she holds back.

Victim – a distressed damsel, she bends, but does not break.

Heroine – a dedicated fighter, she meets commitments.

Caregiver – serene and capable, she nourishes the spirit.

All of these archetypes are accompanied by specific traits, costumes, hairstyles, mannerisms and even background scores. Not to mention that these similar fates often befall these archetypes, whether in a Spanish TV novella or in an Indian family drama. These archetypes are useful to parameterize and track the evolution of characters in a story. When correlated with ratings and audience reactions, they can help recognize traits that work, identifying where characters are on their journeys and where they can be led. There exist similar characterizations of villains as well.

The use of GenAi for these purposes is powerful. Not only can LLMs easily identify, assign and track character archetypes, but they can also generate scripts that incorporate archetypal characters, plotlines, and themes relevant to the script objectives. GenAi-driven scriptwriting tools can assist in crafting compelling narratives that resonate with the target audience.

Leverage GenAi to develop well-rounded and relatable characters that embody archetypal qualities. GenAi can generate character traits, motivations, and conflicts that drive the storyline and engage viewers on an emotional level.

The Neural Alchemy of Storytelling: Decoding Narrative Twists in Screenwriting

The art of screenwriting may be compared to the quest of an alchemist, since it involves transforming the fundamental components of character, place, and conflict into a captivating tale of great value. The essence of this endeavor lies in narrative twists – those unexpected plot developments that seize our attention and imprint stories in our recollection. It is well acknowledged by scholars that there are some characteristics in storytelling that are universal and have a lasting impact across many cultures and time periods. These archetypal storylines form the fundamental basis on which all unexpected turns and surprises are built, serving as the crucial moments within the framework of a novel.

But what makes a good twist? The answer lies at the intersection of cognitive science, emotion, and memory. When a story unfolds in unexpected ways, our brains are jolted into heightened alertness. This is the result of a neurochemical cocktail, spearheaded by dopamine, that rewards the brain's pleasure centers and heightens our engagement. Such neurobiological responses are akin to the workings of a finely tuned engine, propelling the audience through the narrative landscape.

Not all twists are created equal, of course. To be effective, they must adhere to the principles of plausibility and foreshadowing – two narrative constructs essential to maintaining believability and satisfying the audience's craving for coherence. A twist that appears from nowhere may confound viewers, where one that is cleverly seeded throughout the story can lead to elation and a sense of fulfillment. Neuroscience explains that this satisfaction stems from the brain's natural predilection for pattern recognition and problem-solving.

Understanding plot mechanisms can be transformative in the world of television writing. TV shows often rely on high-stakes plots and character development over the course of multiple episodes or seasons, hence the importance of narrative arcs that can sustain viewer interest over time. By mapping out and analyzing the most successful twists from a variety of genres, writers can assemble a "twist library" tailored to the intricacies of long-form storytelling – a veritable Pandora's box of narrative tools at their disposal (Dowd 2015).

The next step in this narrative odyssey is identifying the specific tropes that resonate within particular genres. Whether it's the tragic misunderstandings of romance, the thrilling redemption in action, or the emotional payoffs in family dramas, each genre carries its own rhythm of rising and falling action that speaks to its audience in a unique way. Deft screenwriters learn to navigate these waters, crafting twists that honor the genre's conventions while striving to break new ground.

Empirical data offers yet another clue in this quest. Television ratings provide a quantitative reflection of a show's reach and engagement. By correlating these ratings with particular episodes and their defining twists, writers and producers can gain insights into the effectiveness of different narrative elements. Ratings peaks can indicate moments of collective rapture or shock – signposts of successful dramatic turns.

All of this suggests that while storytelling feels like a purely creative endeavor, it is also one that can be informed and enhanced by rigorous analysis. Machine learning algorithms have begun to play a role in this, parsing through mountains of viewer data alongside narrative structures to detect patterns of success. These data-driven insights pave the way for predictive modeling in screenwriting, where possible plot twists are evaluated not just for their creativity but also for their potential audience impact.

As screenwriting evolves into a data-enriched art form, writers must walk the line between formula and invention. The twist library is a starting point, not a crutch. It's the springboard from which authentic, surprising, and profound stories can leap forth. Through balancing the science of story structure with the art of human creativity, the next generation of screenwriters is poised to enchant audiences with tales as old as time – spun in ways we've never seen before.

Use GenAi to develop complex character arcs that are conducive to unexpected narrative twists. GenAi-driven character development tools can assist in crafting multidimensional characters with hidden motivations or secrets that drive the plot forward.

Employ GenAi to incorporate foreshadowing and subtle clues throughout the script that hint at the upcoming narrative twists. This helps build anticipation and intrigue while maintaining coherence and believability. Infuse the narrative twists with emotional depth and resonance to evoke strong reactions from the audience. GenAi can assist in crafting twists that elicit surprise, suspense, excitement, or even catharsis, depending on the script objectives.

Orchestrating Words, Sounds, and Emotion in Screenwriting Crafting Words that Resonate

Words are the lifeblood of human connection, a conduit for emotion, culture, and identity. Neuroscience reveals that dialogue activates auditory processing networks like the superior temporal gyrus and engages cognitive empathic systems in the prefrontal cortex (Finset 2014). Effective dialogue resonates with the listener by mirroring the rhythms and patterns of genuine human speech. It goes beyond mere information exchange – true dialogue conveys the subtle interplay of intention, emotion, and social dynamics. The science of pragmatics, which explores how context influences the interpretation of meaning, underscores the multidimensional nature of conversations in narratives.

The concoction of compelling dialogue lies in infusing it with the piquant flavors of authentic cultural speech, resulting in a more visceral engagement for the audience. Colloquialisms and localized linguistic idioms invoke the brain's linguistic processing centers and tap into the collective memory embedded within a culture's shared lexicon. When characters use phrases and references that echo the viewer's own lexicon, they close the empathy gap, and identification with the characters deepens.

Sprinkling metaphors into the mix can serve to punch up dialogue and vividly paint mental images. The right metaphor awakens a rush of neural correlates across the brain, from the language-centric left hemisphere to the imaginative right, triggering the angular and supramarginal gyri to process the abstract connections woven by the words (Kim et al. 2022). This figurative language elevates the dialogue from the mundane to the memorable.

The strategic use of cultural anecdotes and examples acts as bridge-building tools in storytelling. These narrative elements encourage the audience's mirror neuron system to fire, a neural network implicated in imitation and empathy.

As viewers relate personal experiences to what is depicted on screen, engagement escalates. Creating dialogue with the dialogue punch-up engine is a form of artistic alchemy. It's a process rich with the potential to turn the base metal of mundane conversation into gold-spun threads of connection. When this alchemy is right, the dialogue pops, and characters transcend the screen, pulling audiences into the heart of the narrative.

The narrative reaches its climax by actively involving the viewers, transforming them from passive spectators into active participants in the unfolding story's conversational rhythm. The conversation punch-up engine possesses a remarkable ability to combine linguistic alchemy and cognitive resonance, resulting in a captivating and enjoyable experience.

Collaborate with GenAi to enhance the humor and wit of dialogue through wordplay, puns, double entendres, and other comedic techniques. GenAi can provide suggestions for punch-ups that add levity and entertainment value to the script.

Use GenAi to refine the timing and delivery of dialogue to maximize comedic impact. Experiment with pacing, pauses, and emphasis to create comedic beats that land effectively with the audience.

GenAi-driven character development tools can assist in crafting dialogue that reflects each character's unique traits, quirks, and motivations. Ensure that dialogue punch-up is culturally relevant and sensitive to diverse audience perspectives and sensibilities. Use GenAi to identify cultural references, trends, and nuances that can be incorporated into the dialogue to resonate with the target audience.

Character Interactions and Audience Appeal

These social exchanges form the backbone of narratives that move us, shaping our perceptions of the characters and their journey. From the hallowed halls of psychology to the cutting-edge laboratories of neuroscience, understanding human interactions has been a pursuit of many disciplines. Modern neuroscience has brought new insights into this understanding through the lens of social cognition – a domain concerned with how people process, store, and apply information about others and social situations. At its core, social cognition studies the neural mechanisms behind how we perceive and decode interactions between individuals.

When characters on screen engage in a relationship, be it conflictual or cooperative, our brains light up in regions responsible for social processing, such as the temporoparietal junction (TPJ) and the medial prefrontal cortex (mPFC) (Tamir et al. 2016). These areas are hotbeds of neural activity that help us infer the mental states of others, a capability known as *theory of mind* (ToM). We intuitively apply our ToM when we witness characters interacting, constantly adjusting our perceptions and attitudes toward those characters. This mentalization affects the deep-seated empathy network within our brains, triggering responses of compassion, admiration, or sometimes, repulsion.

Character interactions in storytelling can be thought of as positive or negative valences that shape the narrative's electric potential. Positive interactions, such as those that exhibit trust, respect, and support, activate regions in the brain related to reward, such as the *ventral striatum* and *ventromedial prefrontal cortex* (vmPFC). These interactions surge through our reward pathways like a heartfelt melody. Conversely, negative interactions stimulate the amygdala and insula, warning us of deceit, conflict, and emotional pain. Whether we are soothed by the harmony of camaraderie or tensed by the dissonance of betrayal, the brain's biochemical symphony plays on.

The weave of interactions between archetypal characters forms a tapestry that is undeniably mesmerizing to the audience. By engaging genres and tropes, writers harness the power of familiar plots with fresh character exchanges to enrich the narrative cloth. Take, for example, the hero's gallant sacrifice or the betrayal of a confidant – these moments pivot not only on the archetypes but on the dynamic interplay between characters. Such dramatic engagements are not just theatrics; they carve neural grooves in our memory, bookmarking the storyline for later rumination and reflecting the audiences' preference in ratings and engagement metrics.

Bringing this insight into the limelight of screenwriting, we recognize the power of systematically analyzing character interactions. Audience ratings can provide a quantitative meter of emotional resonance and satisfaction. Modern techniques such as sentiment analysis and data mining empower screenwriters to decode the formula of success. By assessing which types of relationships and character arcs hit the high notes with viewers, writers can tune their scripts to captivate with an emotional cadence that resonates with the human experience.

Narratives are not set in stone; they metamorphose through interactions, evolving much like the characters within them. It is the unpredictability of these interactions that tread the fine line between cliché and innovation. When a nurturing character reveals a sinister side, or when sworn enemies come together, our brains buzz with the thrill of surprise and intrigue. These moments, orchestrated with finesse, can spiral an ordinary tale into the realm of extraordinary, as the plot thickens with each twist and turn.

Through each disclosure and implementation, the interaction of character dynamics beyond the boundaries of the screen and page resonates throughout the audience's mind. The interplay and drama of the character persists, perpetually captivating and constantly progressing – a testament to the inherent ability inside our essence to establish connections through narratives.

The types of interactions characters typically have in the universe of the story becomes an important framework that can be used to quantify character behavior and thereby search for correlates with audience ratings. In the world of the Indian TV family drama, our research shows that the following types of interactions occur frequently.

Often a character is:

accusing, accused, worried, angered, surprised, rescuing, rescued, panicked, manipulated, manipulating, humiliating, humiliated, escaping, in danger, sacrificing, miserable, trusting, shocked, daring, taking revenge, celebrating, loving, supporting, respected or respecting in their interaction with certain fixed sets of characters. This is certainly not an exhaustive set of possible interactions, but they are nonetheless powerful in being able to construct a web of fixed interactions. Imagine a network connecting all the characters, with pairs of characters being connected by lines that specify what the majority of that interaction is. Immediately complex storylines become easier to parse in these terms, and successful story arcs with high ratings can be correlated with the strong presence of certain structures in this network. It becomes easy then to identify and incorporate specific conflicts and types of resolutions into character interactions to drive the storyline forward and create tension. GenAi can assist in generating dialogue that escalates conflict, resolves disputes, and reinforces key messages. This framework also allows the development of dynamic relationships between characters that evolve over the course of the advertising campaign. GenAi can assist in creating nuanced interactions, power dynamics, and interpersonal dynamics that add depth to the storyline.

In this chapter, we have discussed how to use TV ratings to parametrize plots, archetypes, and character interactions to understand what works and what does not. We have also discussed how to intelligently use GenAi to create newer plots, twists, and dialogues based on the analyzed ratings data. We have also discussed how to have GenAi punch-up dialogues using metaphors as the core linguistic driver.

CHAPTER 18

luxuryGPT – Premiumize Products Using Brain Codes of Luxury

The nonconscious mind has unique structures and systems that enable it to "read and respond" to the cues of luxury.

Knowing them and their functions will enable you to craft everything from packaging to product personalization, and much more.

This information will be especially useful for product messaging.

To that very point: This chapter provides detailed ideas and recommendations on how to apply GenAi in combination with neuroscience learnings for maximum value and effect in marketing.

The Paradox of Plenty: Perceived Luxury in the Economics of More or Less

The perception of luxury is influenced by both scarcity and plenty, making it a slippery slope. The scarcity heuristic and the abundance effect are well-documented concepts in the scientific world, particularly in behavioral economics. Whereas the latter contends that larger numbers can denote luxury for more practical products, the former holds that scarcity raises an item's perceived worth. Therefore, these factors are crucial in determining how consumers view the worth and opulence of goods.

Costco, the American multinational which operates on the principle of bulk shopping, lends an excellent case study to the abundance effect. The size of the products, the scale of the warehouses, and the associated savings all contribute to a sense of accessible luxury for consumers – an unexpected find in the realm of economized shopping experiences. Neuroscience tells us that this sense of gaining more for less can release dopamine, a neurotransmitter that signals pleasure and reward (Zhang 2022), giving shoppers a subtle but measurable buzz.

On the flip side, brands that typically cater to the higher strata exploit the scarcity heuristic. Limited edition releases, exclusive memberships, and personalized services are not just selling the product; they are selling rarity, which triggers a different dopamine response associated with desire and anticipation. It's the hunt for the scarcely available that fuels the luxury fire for high-end consumers.

Traversing from the realms of shopping giants to the intimate space of the laundry room, even mundane products like laundry detergent can be transformed into objects of luxury. Neuroscience research into multisensory perception can explain the appeal of a specially packaged, larger-than-usual detergent bottle. The visual cue of size combined with the touch and even the scent of the product can generate a sensory-rich experience that elevates a routine chore into a luxurious ritual for the budget-conscious.

But how does this transition into a marketing strategy? Understanding these neurological triggers, marketers can create campaigns that offer a sense of abundance or exclusivity, depending on the target demographic's orientation toward luxury. For budget brands, it's about highlighting value and the sensory satisfaction of more: larger quantities, richer experiences, better deals. Injecting this angle with a bit of playfulness – imagine a character joyously swimming in a sea of soft, freshly laundered towels – can resonate well with the consumer's reward system.

Conversely, marketers of premium brands want to curate an aura of exclusivity. Neuroscience aligns with the psychological impact of limited access, urging marketers to capitalize on the thrill of the chase. Here, a more serious tone is usually warranted, as the campaigns could depict discerning customers unlocking private vaults of rare merchandise, or choosing from personally customized options. This aligns with the neural networks linked to decision-making processes, such as those in the prefrontal cortex.

The neuroscience applied to marketing transforms the narrative to target very human responses. When looking at products that are at the lower end of the market, the marketing narrative can afford to be light-hearted and humorous as it celebrates the joy of unexpected bounty (Warren et al. 2018). The vignettes might portray a family's wonder as they discover a larger container of their favorite detergent, a subtle nod to the pleasurable surprise activated in the brain's reward circuitry.

In the realm of higher-end goods, storytelling takes a decidedly more mystique approach. The ads could visually narrate the meticulous care taken in crafting a limited-edition product, highlighting the desire for status and individualism. This not only appeals to the anterior cingulate cortext's (ACC) role in anticipation but also flirts with the dopamine pathways associated with social hierarchy and exclusive access. It is the deliberate unveiling, the whisper of luxury that ignites the cerebral narrative for scarce goods.

Strategic marketing combines the understanding of brain circuits with consumer behavior to tap into the dimension of perceived luxury. Brands must decide whether to fill the sails with the wind of abundance or tack into the allure of scarcity. For the everyday products, campaigns can transform the mundane into moments of delight and indulgence, akin to a welcome plot twist in a familiar tale. In contrast, for the higher echelons of product offerings, the marketing must mimic a carefully composed symphony that crescendos with the reveal of an exclusive product, moving the audience with its elusive notes.

The riddle of why luxury is frequently bottled into diminutive containers or numerically limited releases can be decoded through an exploration of our brain's reward system. The lure of small, curated selections lies in its titillation of our dopamine pathways, tuning the mind to resonate with the concept of rarity.

A curious case to examine is the packaging of perfumes. The vessels that hold aromatic liquids are designed to evoke an aura of exclusivity (Horoszko et al. 2018). Even without the scent reaching the nostrils, the mere sight and feel of a delicately crafted bottle promise an opulent experience. Visual processing in the brain's occipito-temporal cortex, coupled with the expected pleasure encoded in the orbitofrontal cortex, sets the stage for a perception of luxury well before the actual product is used.

Amplifying this concept, vintners and watchmakers trumpet their items as limited editions. This creates a sense of urgency and value that is amplified by the finite availability of these luxurious goods. When connoisseurs hear of a vintage with only a thousand bottles in existence, or a watch series capped at a few dozen pieces, their nucleus accumbens – often referred to as the brain's pleasure center – lights up, transcending the object from mere material to a covetable treasure.

This alchemy of scarcity is not accidental but deeply intertwined with historical associations of preciousness. Gold, diamonds, and other rare-earth elements have always been celebrated not only for their intrinsic properties but also for their rarity. The human brain has developed a fascination for elements that stand out in scarcity, relating sparse availability directly to worth, a concept solidified in various fields including biological market theory (Srivastava et al. 2001).

The paradox then takes a plot twist when it comes to marketing strategies across different price points. For more economically accessible products, the approach swings toward providing more for less, creating a perception of bounty. Contrastingly, at the apex of the price pyramid, luxury is engineered through the psychology and neurobiology of less-is-more, quantitatively constricting supply to magnify desire and perceived value.

A delightful spin on this narrative is the creation of "variety packs," which can transform even the common chocolate box into a veritable treasure trove. Despite the chocolates' collective smallness, the brain's ventral tegmental area (VTA), responsible for processing novelty (Krebs et al. 2011), creates a dynamic sensory and emotional palette that feels both novel and luxurious. This multifaceted approach to luxury marketing tickles a different part of the brain – the desire for variety and complexity, akin to an intellectual tease.

No matter the direction, the fusion of humor, science, and storytelling can elevate a product's market position. For the consumer navigating the aisles of a megastore or the pages of a luxury catalog, it is the story, the promise of delight or distinction, that endears the product to the heart and excites the brain. The paradox of plenty and the allure of less becomes the marketer's narrative canvas, ready to be painted with the psychological and neural hues of luxury.

The paradoxical nature of luxury, teetering between plenty and paucity, guides the deft hands of marketers. Using the brush of neuroscience, they can paint a universe where every product, regardless of its niche, glistens with the varnish of luxury to some degree. Be it the boisterous bounty at Costco or the exclusive essence of a first-edition timepiece, the underlying science remains resolutely the same, only its application in the theater of markets fluctuates. That is the artistry of marketing, a realm where perception is the ultimate reality.

The Temporal Tapestry: Time as a Weave of Luxury

In the sphere of luxury, time is an architect, shaping objects not by form, but by history, process, and patience. The connotation of time as a conduit of value finds its roots in our psychological makeup, embedding the idea that temporal investment equates to intrinsic worth. This is an interplay of chronology and quality, where the time-tale told by a brand can transform perceptions and pivot the ordinary into the realm of desirability and grandeur.

The crafting of a fine wine or whisky is an oft-trotted example of temporal allure in luxury marketing. The aging process, often taking decades, is not merely a chemical necessity but serves as a narrative. The consumer's mind, sparked by cues of extensive maturation, activates regions associated with anticipation and expectation, even before the first sip. The concept of prolonged creation is enough to imbue the notion of a rich, complex flavor profile, warranting a higher pedestal and price point.

Neuroscientifically, the visualization of time-intensive processes can stimulate the reward circuits in the brain such as the ventral striatum, lighting up at the promise of something extraordinary (Hamid et al. 2016).

These processes are orchestrated by a ballet of neurotransmitters where dopamine plays the conductor, influencing how we perceive and assign value based on time.

In the realm of goods beyond beverages, craftsmanship that touts meticulous handiwork over countless hours triggers a similar neuronal response. The consumer's prefrontal cortex, tasked with planning and complex behaviors, rationalizes that the painstaking detail and slowed creation mandate reverence and reward. It is in these moments, when strokes of time are painted on the canvas of a product, that luxury is not only seen but felt.

A lesser-known angle of temporality in luxury is the anticipation it fosters. The wait for a product, whether due to the length of its construction or the delay before acquisition, can enhance desirability through the mechanism of delayed gratification. The interplay of time in this manner satiates a deeper, more cerebral craving for fulfillment, embedding luxury within the mental vaults of pleasure and patience.

Marketers, like craftsmen of desire, thus use the dimension of time to weave narratives that charm the basal ganglia, home to the brain's reward system. They spin yarns of time-transfixed artisans, their dedicated hours contributing to an almost mythological birth of a luxury item. Such storied detail can be light-hearted – a watchmaker jocularly bemoaning the near-eternity it takes to perfect a single timepiece – yet it carries the punchline of prestige.

As science seamlessly transitions into the practical, one finds that the role of time in the lore of luxury is not rigid. It's a fluid saga where even contemporary products entice through "limited drops" that accentuate the brevity of availability, contrarily equating the quick snap of time with value due to exclusivity.

In conclusion, through the neural and the narrative, time emerges as an invisible yet potent ingredient in the concoction of luxury. By recognizing the cerebral choreography that time incites, one can appreciate its indelible impact on constructing what is coveted. The luxury connoisseur's brain applauds not the swiftness of creation but the longevity of craft, the patient ferment of ideas into irresistible allure.

GenAi can help create visuals and narratives that evoke a sense of timeless elegance, emphasizing the enduring appeal and quality of luxury products or experiences. Highlight the meticulous craftsmanship and rich heritage behind luxury brands by using GenAi to develop storytelling elements that showcase the time and dedication invested in creating each piece.

The Illusion of Exclusivity: Unavailability as the Fourth Dimension of Luxury

In the realm of luxury, unavailability is a subtle yet persuasive magician. It conjures a mirage of opulence and distinction, through the sheer force of scarcity. The neuroscience underpinning this phenomenon is rooted in the brain's response to rare stimuli. The mesolimbic pathway, particularly the nucleus accumbens, acts as our reward center, and it is known to ignite with anticipation and pleasure when faced with the prospect of obtaining something exclusive. This pathway doses the brain with dopamine when the elusive becomes attainable, making the chase for the unattainable an addictive proposition (Sjostrom et al. 2016).

With Tesla's Cybertruck, for example, the allure of its futuristic design is rendered even more seductive by the wait it demands. The lengthening queue morphs into a status symbol, transforming the electric marvel into a luxury icon. Anticipation of reward is a key pleasure driver in the brain and the deferred gratification, linked with waiting increases the desire for these prized possessions. The dopamine rush of finally obtaining the coveted object turns ownership into a triumph, an emotional victory that is both deeply personal and highly brag-worthy.

Similarly, the Leica Q3 camera's allure is intertwined with its unreachability. This unavailability triggers an intrinsic fear of missing out (FOMO), which is a psychological pattern underpinned by the amygdala (Tanhan et al. 2022). FOMO energizes consumers to pursue scarce commodities, magnifying their perceived worth and the personal significance of their acquisition.

The perception of scarcity as a hallmark of luxury is not a modern invention but rather an age-old response ingrained in human cognitive evolution. Resource scarcity played a significant role in survival, making the brain adept at elevating the value of what is hard to come by. This primordial bias is harnessed by marketers to present products as limited editions, conferring upon them an aura of immediate premiumization.

Transitioning the science into practice, marketers adeptly sculpt their strategies around the brain's proclivities. By presenting their wares as one-of-a-few, such as a limited print run of 20,000, the ordinary metamorphoses into the extraordinary. They employ principles like perceived loss, whereby the potential inability to obtain a product increases its perceived value – a psychological effect well-documented by loss aversion theories (Kahneman & Tversky 2013).

The alchemy of these marketing maneuvers is where the nitty-gritty of neuroscience garbs itself in the shimmering attire of storytelling. As much as our brains love a solid fact, they're equally beguiled by a charming yarn. Marketers create a perception of perceived exclusivity by using a deeper understanding of the brain's desire for the uncommon and the unachievable to elevate things to the pinnacle of desirability. The next time you want something that is out of reach, keep in mind that your brain is tricking you into believing it to be exclusive.

Emphasize the exclusivity and rarity of luxury goods by leveraging GenAi to craft narratives that underscore the limited availability and unique craftsmanship that make them highly coveted. Utilize GenAi to craft narratives and visuals that highlight limited-edition releases of luxury products, emphasizing their exclusivity and desirability. GenAi can help create compelling storytelling elements that communicate the exclusivity of owning a rare item.

Rarity Refined: The Dazzle of the Uncommon in a World of the Common

Neuroscience tells us that rarity triggers a response in the brain's reward circuitry, similar to that which occurs when we receive unexpected positive stimuli. The ventral striatum, in particular, becomes activated when we encounter rare items, triggering a release of dopamine, a neurotransmitter associated with pleasure and desire. This response to rarity is not coincidental but deeply rooted in our survival instincts, where unique resources could convey significant advantages.

These resources need not be something as grand as precious metals or gems; even culinary ingredients like the famed truffle or the gold foil–coated confections signal luxury due to their scarcity. The reason such items demand attention is that rarity not only connotes a certain uniqueness but also speaks of the extreme effort and singularity required to attain them. Rarity thereby becomes synonymous with luxury through the psychological association of effort with value (Inzlicht et al. 2018).

Further affirming the science, research reveals that when we label an item as "rare," even if the consumers are unaware of its utility, they are likely to rate it as more valuable than its common counterparts. This perceived value is amplified if the rare item can be integrated into a personal or social identity, making it not just an object of desire but a vessel of self-expression and social status.

The extraordinary occurrence of biologically limited events, such as an annual salmon run, transforms these ingredients into curiosities. The neurobiological systems engaged here are those governing the natural calendar, ethologically primed to prioritize seasonal and circadian rhythms.

But it's not just the olfactory luxuries that play this tune; consider the realm of spirits with ingredients that reach mythic status due to their scarcity. The mere presence of gold flakes in a liqueur, irrespective of their nominal utility, sets off the brain's associative learning processes, harking back to the age-old equation of gold with power and wealth. Marketers leverage this, tapping into the basal ganglia's reward systems with campaigns that showcase the material's luxury through its rarity.

The idea of rarity doesn't need to wrap itself in seriousness. Humor can enrich the appeal, as illustrated by marketing campaigns infusing the exclusive with the mundane. Imagine an artisan bread loaf so unique it has its own comedic social media account, lamenting its lonely existence as "the single bachelor of the bakery." Such levity engages the cognitive recognition of value in a manner that's both playful and irresistible to the brain's reward pathways.

Rarity dances with our ancient instincts and modern aspirations alike, enchanting us into coveting the not-so-prolific pleasures. It's in this treasure hunt, led by both science and sensibility, that we find the zestful zenith of luxury.

Thus, the next time a rare truffle appears on your plate, or a limited-time offer lands in your inbox, remember the complex neural tapestry that weaves rarity and luxury into a single, captivating narrative. It's this interplay of scarcity, storytelling, and the science of desirability that spiritedly captions the human journey from the ordinary to the extraordinary – a journey abound with sly winks and knowing nods to the rare joys of a world replete with bounty.

Provide consumers with behind-the-scenes glimpses into the craftsmanship and creation process of luxury goods. GenAi can assist in generating content that showcases the meticulous attention to detail and artisanal techniques that contribute to the rarity of luxury products.

The Handcrafted Halo: Crafting the Neuroaesthetics of Luxury

The human touch holds a remarkably esteemed position in the kingdom of luxury. Across cultures and epochs, the act of creation by hand has been lauded, infusing crafted items with an aura of exclusivity and charm. Neuroscience reveals that our preference for handcrafted items may stem from the brain's reward system, particularly the nucleus accumbens, where a surge of dopamine appreciation flows when we recognize the skill and labor of artisanal work. This suggests a deep-seated neural reverence for human craftsmanship as juxtaposed against machine-like precision.

Craftsmanship is often related to mastery and skill, concepts that are celebrated within the framework of the prefrontal cortex, the part of the brain associated with planning and complex cognitive behavior. The cerebral celebration of these intricate creations may also be influenced by mirror neurons, which respond not only to action but to the perception of action, essentially allowing us to appreciate the artisan's expertise vicariously.

Adding another layer to this neurological tapestry, the concept of *flow*, wherein an individual is fully immersed in a feeling of energized focus, full involvement, and enjoyment while creating something, likely contributes to the humanistic allure. Observers can intuit the mental state of artisans during creation, heightening the perceived value and luxury of handcrafted goods.

But why the strong inclination toward handmade items? Anthropological studies suggest that toolmaking and the manual dexterity tied to crafting items by hand were pivotal in the cognitive evolution of humans, fostering intricate social structures and cultural development. This historical significance likely resonates within our evolved neural networks, granting handcrafted items an esteemed status that resonates across the subconscious expanse.

The science of neuroaesthetics offers insights into why we find handcrafted items to be especially alluring (Frater et al. 2018). This field of study investigates the neural and biological foundations of aesthetic experiences and integrates sensory, motor, and emotional reactions to the perception of art and beautiful objects. It explains why we might appreciate the irregularities and textures that come from something lovingly made by an artisan's hand, perceived through the activation of the insular cortex, which is involved in emotional salience.

Shifting focus from the neural mechanisms to practical applications, brands involved in the luxury sphere often highlight the handcrafted nature of their goods. This emphasis taps into consumers' evolved appreciation for craftsmanship, allowing them to weave a seductive thread of storytelling around their products. The premium teapot isn't just a vessel for tea; it's a relic of immersive labor and traditional skills passed down through generations.

Consider the satisfaction derived from sipping a "handpicked" artisanal coffee – beyond the taste, it's the lore of hands sorting beans, the whisper of intimate selection, that warms the soul. This approach not only informs consumers but also strikes a chord with their inner humanness and yearning for connection.

Create visually stunning content that showcases the intricate craftsmanship and unique details of handcrafted luxury items. GenAi can assist in generating high-quality imagery, videos, and animations that highlight the craftsmanship process and the beauty of the finished product.

Profile individual artisans and craftsmen who contribute to the creation of handcrafted luxury products. GenAi can help develop compelling profiles and interviews that showcase their expertise, passion, and dedication to their craft, adding a human touch to the advertising campaign.

Tailored Elegance and the Neuroscience of Personalization

Personalized products inhabit a special echelon in the world of luxury, stroking the strings of exclusivity and identity. When we delve into the neuroscientific intricacies of this phenomenon, we find that the concept of having something "made just for you" caters to our deep-seated neural craving for uniqueness. The sense of self is reflected in brain regions such as the medial prefrontal cortex (MPFC), which lights up in response to stimuli that align closely with our personal narrative and self-concept. Custom-made objects mirror our persona, leading to an amplified sense of luxury.

The psychology of ownership and its correlation with an object's value is well-documented in the endowment effect, whereby people ascribe more value to things merely because they own them. When that ownership is defined by customization, the object moves beyond possession to become a statement of the self, intertwining with our identity and consequently increasing its perceived worth.

Bespoke services cater to our inherent need for control and agency. Neuroscience has shown that having a sense of control over one's environment activates the dorsolateral prefrontal cortex (DLPFC), a key area in controlling executive functions and decision-making. A bespoke item, by virtue of its tailored nature, offers control over the final outcome, therefore escalating its emotional and cerebral valuation.

Delving further into the tailored experience, neuroimaging studies reveal that anticipation plays a substantial role in enhancing our enjoyment of end products. The act of customizing – selecting specific elements to suit personal preferences – triggers the nucleus accumbens, signaling an anticipatory pleasure response. This neural reward circuitry primes us for a heightened level of enjoyment when we receive the finished personalized product.

On a molecular level, the neurotransmitter dopamine, heavily implicated in the brain's reward system, propagates the feelings of pleasure associated with expectation and accomplishment. When the process of designing or selecting customization options for a personal product takes place, the surge of dopamine sets a positive emotional foundation for the bespoke experience, cultivating an exclusive mystique around the finished piece.

The oxytocin system further enriches this narrative, as it is implicated in social bonding and trust. When a product is crafted to one's specifications, a connection is formed not just with the item but also with the creator. This injects the unique dynamic of relationship into the perception of the product, imbuing it with a lore and significance that off-the-shelf items simply can't replicate.

Shifting to market applications, modern brand narratives thrive on the allure of personalization. They tap into consumers' intrinsic motivations by offering "your very own" version of a product. Whether it is a monogrammed leather bag or a car in a personally chosen hue, these offerings resonate on a level that surpasses mere ownership; they represent individual influence and creative partnership.

The process of crafting a bespoke experience resonates deeply within our brain's architecture, particularly through the concept of *autonoetic conscious-ness*. This form of consciousness enables us to mentally place ourselves in the future, imagine personal events, and craft experiences (Klein 2016). Autonoetic consciousness lies at the core of creating personalized experiences, whether it be a build-your-own salad or a freestyle beverage. By projecting our preferences into tangible outcomes, the brain beholds such customizations as a reflection of the self, giving rise to a perceived sense of luxury.

Highlight the opportunity for customization and personalization that hand-crafted luxury products offer. Use GenAi to create interactive experiences that allow consumers to tailor their purchases to their unique preferences, reinforcing the exclusivity and individuality of each piece.

Emphasize the rarity and exclusivity of handcrafted luxury items through limited-edition releases and special collections. GenAi can assist in crafting mes-saging and visuals that communicate the limited availability of these unique pieces, driving demand among consumers.

Chromal Elegance: The Spectral Alchemy of Luxury

Color speaks a silent, yet resoundingly rich language, one that luxury has mas-tered fluently. The science behind color perception begins in the eyes, where pho-toreceptors translate light into neural signals. These signals journey to the brain, engaging the visual cortex as well as a cascade of processing that imbues color with emotion and meaning. This neuro-chromatic dialogue crafts the palette of what we perceive as luxury.

Gold, the timeless symbol of wealth and opulence, elicits its luxury not merely through association, but through its stimulation of the reward circuitry in the brain. The sight of gold activates the brain's visual pathways in a distinct manner and engages regions associated with reward processing, such as the orbi-tofrontal cortex, emphasizing its perceived value and rarity.

Contrast is key in creating visual impact, and the combination of black and gold is neuroaesthetics in action. Black, rich in simplicity and depth, when juxta-posed with the lustrous sheen of gold, produces a visual frequency that resonates with the notion of sophistication and elegance. This stark contrast elicits strong neural responses, enhancing the luxurious aura of the pairing.

The classic allure of the black and red palette, often sported by high-end auto-mobiles, draws its luxury from a blend of power and passion. Red stands as a color of heightened emotion and intensity, and studies indicate it can increase physiological arousal, possibly linking it with excitement and speed, characteris-tics celebrated in luxury performance vehicles.

For the feminine aesthetic, the shimmer of sheer fabrics adds a layer of tactile luxury. Sensory integration in the brain, particularly within the insular cortex, is attuned to combining visual and tactile stimuli – an interwoven play of sheen and texture that can translate into an enhanced perception of refinement.

Cultural factors entwine with neural wiring to inform color associations. White, often linked with purity and clarity in many societies, carries an air of refined simplicity that's prized in luxury spaces. When this cultural symbol is blended with the natural appeal of the color red, which is frequently associated with love and vitality, it cultivates a luxury spectrum well suited for opulent fashion.

Translating from the neurons to the concrete, luxury brands habitually infuse their products with the chromatic codes of richness (Som & Blanckaert 2015). A swathe of gold or a brush of black on a product's finish can alter consumer perception, nudging the mental scales toward an appraisal of luxury. This strategic application of color is grounded not only in aesthetic tradition but also in the modern understanding of neuroscience.

Generate visually stunning imagery and graphics that showcase luxury products using elegant color schemes. GenAi can assist in creating high-quality visuals, such as product photos, advertisements, and social media content, that highlight the elegance and refinement of the brand.

Create atmospheric lighting effects that complement elegant color schemes in advertising campaigns. GenAi can provide recommendations for lighting techniques and effects that enhance the visual appeal and ambiance of luxury settings and environments.

The Gravity of Luxury: Weight and Stature in Perceived Value

In the realm of luxury, an object's physical weight carries a surprising psychological heft. Our perceptions of weight and value are intrinsically linked, a coupling rooted in the brain's somatosensory cortex – a region responsible for interpreting tactile feedback from our environment. The perception of weight engages a set of expectations regarding the quality and worth of an item, where heavier feels synonymous with better.

Heavy objects trigger a multifaceted response in our cognitive architecture. According to embodied cognition theories, physical experiences can shape cognitive processes. This means that the act of holding something substantial can invoke concepts of importance and value, shaping our judgment far beyond its actual functionality – a pleasing quirk of human perception that high-end winemakers have tapped into with their sumptuously ponderous bottles.

Moreover, the "gravitas" of an object's weight is not limited to tactile sensation; it reverberates deep within our cognitive biases. Psychologically, heaviness is associated with seriousness and reliability – a sentiment that dates back to archaic times, where the worth of an object could be gauged by its density. The dense bottle thus becomes a modern-day talisman of luxury and quality.

Conversely, height and elevation are inherently perceived as indicators of grandeur and prestige. A towering structure – whether a skyscraper or product packaging – elicits a psychological response signifying dominance and superiority. In luxury, this vertical advantage is seen as emblematic of a higher standing, quite literally, in the roster of desirables.

At the crossroads of stature and neuroscience, cognitive mechanisms like perceptual symbolism come into play. In the brain, perceiving an object's height not only involves spatial awareness but also transfers to metaphorical thinking about power and status. Therefore, a tall and imposing product can command our reverence in the same way majestic mountains do – an ode to verticality steeped in the neural perception of prominence.

When it comes to compact luxury, the paradox unfolds. Smaller items, particularly when scarce or crafted with meticulous detail, evoke a sense of exclusivity. Just as a jeweler's loupe magnifies the craftsmanship of a tiny brilliant-cut diamond, so does our brain's attention to detail spotlight the understated elegance of the minuscule. The meticulousness required to craft something minute maps onto neural substrates charged with processing fine motor skills and attention.

Marketers cleverly highlight the solidity of a product or its lofty presence, drawing forth the substantial from the substance. They build towers of desire from the base materials, ingeniously architecting campaigns to echo the heft or height of their offerings.

When considering the heft and height of luxury, we recognize that weight and stature play significant roles in value perception. Bridging the sensory and the cerebral, luxury brands wield the dimensions of weight and stature to tantalize neural circuits so that what feels heavy in hand and stands tall to the eye translates to a symbol of unparalleled quality. Whether it's the density of a wine bottle or the height of haute couture, luxury, in essence, is as much about the brain as it is about brawn.

Showcase luxury products in a way that emphasizes their weight and physical presence. Use GenAi to design product displays, packaging, and promotional materials that highlight the substantial feel of the product, creating an impression of luxury and quality.

Emphasize the use of premium materials and finishes that contribute to the heft and weight of luxury products. GenAi can provide recommendations for materials that convey a sense of solidity and durability, enhancing the perceived value and luxury of the product.

The Unraveling Rapture: Ribbons as the Threads of Opulence

The ribbon, a simple strand of fabric, invokes a profound neurological journey from anticipation to delight. Ribbons adorn the most sought-after luxuries, signaling more than an ornamental finish; they symbolize the anticipation of a hidden treasure. Neuroscientifically speaking, the process of unraveling a ribbon activates the brain's reward system, involving key areas such as the nucleus accumbens and the ventral tegmental area (VTA), which are flooded with dopamine during anticipation of a rewarding experience. This anticipation enhances the perceived value of the concealed gift, with every twist and bow meticulously crafted to heighten the senses.

The action of untying a ribbon calls on the brain's somatosensory system, stimulating a tactile interaction that touches on neural substrates of touch and manual manipulation. Tactile engagement is a potent activator of the insular cortex that translates sensory experiences into personal emotional relevance. Here lies the auxiliary charm of the ribbon; it is not merely seen, but felt, unfolding the sensory prelude to luxury with a gentle pull.

Neuroaesthetics – the cognitive study of beauty and art – reveals how visual cues like the soft sheen and elegant curvature of ribbons can sway our preferences. The sight of satin ribbons evokes a perception of smoothness and premiumness, engaging visual processing areas such as the occipital lobe. The careful placement of a ribbon transforms ordinary packaging into a visual and tactile invitation to splendor.

Evolutionarily, human beings have developed keen pattern recognition abilities, making us attuned to the ornamental wrapping that signifies care and consideration (McNeil & Riello 2016). According to evolutionary psychologists, such finery could trigger our natural predisposition to seek out high-quality, nurtured resources, enhancing the impression of the gift. Consequently, ribbons have become shorthand for meticulously selected and prepared treasures.

The concept of gifting extends to realms beyond physical presents; it aligns with the deep-seated human tradition of reciprocity and social bonding. Social neuroscientists note that gift-giving can instigate the release of oxytocin, the "bonding hormone," which fosters a sense of connection and trust between individuals. The ribboned package is thus a conduit for emotional exchange, a nonverbal communicator of social ties and appreciation.

Clever advertisements employ the allure of ribbons to suggest a personal encounter. They invite viewers to become participants in an unfolding story, conjuring the sense that what lies within is bespoke and exclusive. A commercial might highlight the "unwrapping" of a new car model, the shiny ribbon morphing into the road it glides upon – a subtle but memorable fusion of product and presentation.

As we tie up our exploration of luxury ribbons, we recognize their strength as both symbol and sensory stimulant. Ribbons harness a range of brain systems – visual, tactile, and emotional – transforming the act of unveiling into a celebration of the senses. Next time a velvet ribbon is drawn away to unveil the splendor beneath, remember the neural symphony playing in tandem with that simple gesture – a grand opening conducted by the mind itself.

Showcase luxury products being adorned with ribbons in gift-wrapping imagery. GenAi can assist in generating visuals that depict the unboxing experience, highlighting the luxurious presentation and attention to detail that ribbons add to the overall packaging. GenAi can also assist in generating creative concepts and visual elements that leverage the symbolism of ribbons to communicate themes such as celebration, luxury, and exclusivity.

Sensual Shadows: The Neuroscience of Glamorous Luxury

Dark sensuality and glamour, often manifested in serene candlelight and the allure of scattered petals, weave a tapestry of luxury. The intricate interplay of light and shadow produced by flickering candle flames captivates our visual cortex, drawing us into a world of intrigue and intimacy. This dance of luminescence and obscurity tantalizes neural pathways, evoking a sense of mystery and coaxing a luxurious feel from the simplest of settings.

Our proclivity to associate ambient, moody lighting with luxury has deep neurological roots. The hormone melatonin, which regulates sleep and wakefulness, rises in low-light environments, potentially enhancing moods and creating a comfortable setting for indulgence. The sensual interplay of shadow and light, especially in the context of luxury, can stimulate an environment conducive to relaxation and elevated experience.

The sense of touch, too, is keenly engaged amidst the opulence of strewn petals, soft to the skin and pleasing to the eye. The sensation processed by the brain's somatosensory regions can translate mere physical contact into an emotional narrative, one of decadence and care – a tactile whisper evoking the grandeur of life's finer moments.

The association of specific aromatic experiences with luxury taps into our olfactory system, where scents trigger profound responses within the limbic system. The fragrance of a candle or fresh petals can conjure memories and feelings of previous luxurious encounters, reinforcing the neural linkage to upscale experiences.

Underlying the psychology of luxury is the theory of constructed emotions, which suggests that our brains create emotions from a combination of sensory input and past experiences. Thus, the traditional symbols of luxury – candles and petals – activate neural constructions of pleasure and affluence, aligning physical stimuli with an internal sense of opulence.

Moving from the realm of scientific inquiry to practical application, luxury marketing harnesses sensory cues to evoke emotions, crafting a scene where each element, from lighting to texture, builds layers of a luxurious narrative. Through the neuroscience of engagement, we learn that an ambiance replete with warmth and lures consumers into a lived story of allure, creating an aspirational tableau with synaptic elegance.

The bridge between stark science and everyday splendor is crossed with ease. Vacillating shadows cast by candles against the walls of a spa invite patrons to let go of the prosaic and enter into refinement and allure – a well-orchestrated serenade of the senses, where luxury is as much about psychology as it is about setting.

The light-hearted facets of glamour shine through in creative visualizations catering to a market that delights in luxurious escapism. Imagine a boutique hotel's advertisement where rooms are "dressed in shadows and whispers," promising an experience that's more than a stay – it's a sensory sonnet.

As we close the curtain on the sensual embrace of glamorous luxury, we acknowledge the potency of darkness and desire, both wrapped in sensory stimuli that delight the mind and body. Through understanding the neuroscience behind these luxurious codes, we demystify how our brains are primed to indulge in the lush symphony of shadowed eloquence. The next time candlelight graces your table, remember the neural notes it plays, casting a grandeur that transcends the candle's wick and wax.

Use sensual shadows to accentuate the contours and features of luxury products in advertising campaigns. GenAi can help design lighting effects and shadow play that draw attention to key product attributes, creating a sense of depth and dimensionality.

Create atmospheric ambiance using sensual shadows to evoke a mood of luxury and refinement in advertising campaigns. GenAi can assist in generating lighting setups and shadow effects that enhance the visual appeal and emotional impact of the campaign, setting the tone for a sensual and immersive experience.

GenAi can effectively utilize the cues and clues of how luxury is perceived by the brain to generate: messaging, imagery, packaging, and advertising. This chapter has covered the core methods to be algorithmically unleashed to create the perception of luxury.

CHAPTER 19

buzzGPT – Drive Virality through neuroAi

In this age of social media, one of the most highly praised and sought-after phenomena is "going viral." GenAi has the algorithmic power to create viral content by selectively embedding the techniques outlined in the chapter.

The multiplier effect can be enormous.

If your brand is lucky enough to win in this rarefied online lottery, buckle up. It's going to be one heck of a ride.

But creating, or even forecasting, posts that go viral remains in the realm of the unknown.

Until now.

In this chapter, neuroscience pulls back the curtain on how and why online material achieves escape velocity and explodes into virality. You will learn the mechanisms in the mind that reward material the viral crown. You will understand the specific key elements that help drive virality.

When you finish this chapter, you will have the knowledge and the tools to significantly improve your brand's chances of breaking out.

Unique Motion, Novelty, Error, and Ambiguity

Virality is the dream of every marketer. Content is so compelling that you just have to share it – content that boosts the perception of oneself to our peers or to those that matter. It is abundantly clear that virality is about self, and the emotions we can spread.

How exactly does one tell a GenAi algorithm, "Give me content so awesome that humans will want to share it, talk about it, create talkability with it"? *Talkability* has to do with extremes, and that makes things go viral.

We begin by understanding the core of what makes content viral. There are principles of neuroscience that "compel" us to share.

Neuroscience has delineated the *cognitive and emotional processes* that underlie our attraction to motion, novelty, error, and ambiguity.

Motion captures our attention because of our evolutionary history, where understanding and responding to movement was a matter of survival. The superior colliculus, a structure in the midbrain, is specialized for detecting and orienting to movement. Not all motion creates virality – but unique motions arrest the brain, slow motion arrests the brain, and motion that conjures messages arrests the brain.

Novelty, in turn, engages the brain in a different way. The hippocampus, associated with memory, shows increased activity when processing novel stimuli. The release of dopamine during novel experiences has been noted to make these experiences feel rewarding, heightening the likelihood of memory formation and sharing. The emphasis of novelty is on the unexpected, and that creates virality. Introducing novelty in marketing campaigns can be highly effective. Incorporating dopamine-driven learning processes, novel marketing endeavors create memorable and shareable experiences, linking the uniqueness of the content to a product or brand in the consumer's memory.

Errors have a unique place in drawing attention and desire for dissemination. The anterior cingulate cortex (ACC) is well-studied for its role in error-detection. When we notice an error, there's a degree of emotional arousal affiliated with the unexpected, which is a potent driver for social sharing. Errors create particularly viral moments if committed by someone known or a celebrity. Utilizing recognizable errors in content as a hook, intentionally or as a serendipitous opportunity, can lead to heightened sharing. When people spot an error, the urge to correct and discuss it can increase content virality, making it beneficial to create a narrative around such occurrences.

Ambiguity engages multiple areas, including the amygdala, which processes emotions and uncertainty. Ambiguous stimuli often lead to increased curiosity and discussion as individuals seek to resolve their uncertainty by sharing and receiving input from others (Gottlieb & Oudeyer 2018). Creating content with elements of ambiguity lures the audience into a deeper engagement. Ambiguous advertising can ignite imagination and debate among the audience, prompting them to seek out more information or resolve their curiosity through sharing and discussion.

Exploring how these concepts are integral to contemporary content sharing, marketers and content creators capitalize on extremes in media. Stimulation of the superior colliculus not only garners attention but may elevate physiological arousal, sparking conversations around the shared content.

In transitioning to a more applied and understandable presentation, it becomes apparent that by exploiting these neurological tendencies, content creators can craft compelling stories that stick. Think of an ad that starts with a high-speed chase, introduces a novel gadget, exhibits a humorous "error," and culminates with an ambiguous ending – the brain's reward system practically lights up, itching to hit share.

To sum up, the intersection of neuroscience and digital sharing showcases the beauty of the human mind when engaged with extremes of motion, novelty, error, and ambiguity. By leveraging these elements, marketers create a playground for the brain, enticing the sharing of the experience with others, thereby turning viewers into willing and active participants in the storytelling process, especially when it is above the normal threshold and a bit extreme.

Slow Motion Sensation: Virality in the Temporal Dimension

In the realm of visual perception, time is a critical factor. The human brain is wired to respond to motion, and how that motion is presented can amplify this response. A study by VanRullen et al. (2005) found that the human visual system is particularly sensitive to changes in speed, which includes the deceleration inherent in slow-motion footage. Such sensitivity to temporal variations plays a key role in how we engage with visual content.

Slow motion has the distinct power to magnify an experience, allowing viewers to absorb details normally missed at regular speed (Asamoah et al. 2016). This phenomenon can be attributed to the way our brain processes sensory information. Bilateral regions of the cerebral cortex are involved in motion perception, including areas MT and MST, which are specialized for analyzing motion dynamics.

By altering the temporal dynamics of video content, creators capture viewers' attention and heighten emotional responses. This enhancement is not just an artistic choice; it has a basis in neuroscience. The experience of perceiving time as dilated during intense emotional situations is reminiscent of the slow-motion effect. This temporal distortion can occur during adrenaline-rich moments, which increase arousal and memory encoding.

From a marketing perspective, incorporating slow motion into content can make certain moments more memorable and share-worthy. When viewers witness a moment that is viscerally impactful, such as a slow-motion sequence, it may activate a form of empathetic resonance with the content, making them more likely to share it with others.

In the transition to a lighter style, let's imagine the typical "epic fail" video. Why do these clips tend to go viral when slowed down? The brain becomes engrossed in the anticipation and narrative constructed from prolonged actions. At the same time, slow motion affords the viewer the luxury of catching every humorous nuance that's easily missed at full speed.

Moreover, slow motion can play into the humorous juxtaposition of speed. A sloth on fast forward? That's amusing because it defies expectation. A sprinter in slow mo? Equally delightful, as it lends gravitas to mere milliseconds. This play on speed extends to marketing, where a product's quick action might be slowed to emphasize efficiency or effectiveness, creating an almost comedic contrast with the reality of hustle culture.

Approaching application, the deliberate use of slow motion in advertising and social media content can create a "moment" out of the mundane. By taking the trivial and stretching it out, marketers can emphasize aspects of a product or narrative that might otherwise go unnoticed but resonate with the viewer on a slower, deeper level.

As we wrap up this temporal journey, it's evident that slow motion isn't just a nifty trick; it leverages our neural circuitry in a way that can evoke laughter, empathy, and awe. Strategically sprinkled throughout content, it becomes a catalyst for virality, cutting through the endless stream of media to deliver something almost paradoxically swift in its ability to capture our attention and persist in our memory – all by taking its time.

Generate visually stunning slow-motion footage using GenAi to showcase product features, details, or actions in a captivating manner. Slow-motion sequences can highlight the elegance, craftsmanship, and quality of luxury products, captivating the audience and encouraging shares.

Use slow motion to evoke emotion and create a deeper connection with consumers in advertising campaigns. GenAi can assist in generating slow-motion scenes that convey feelings of awe, excitement, or nostalgia, resonating with viewers on a visceral level and prompting them to share the content with others. GenAi can help create slow-motion sequences that highlight key features or functionalities, allowing consumers to see the product in action in a mesmerizing way.

Charismatic Circuitry: The Neuroscience Behind Why We Love "Almost Human" Viral Sensations

The propensity for anthropomorphism, or the attribution of human characteristics to nonhuman entities, is wired into our brains. It's a byproduct of our social cognition evolutionarily designed to understand and anticipate others' behaviors. This phenomenon activates a network in the brain known as the *social brain,* including areas like the medial prefrontal cortex (mPFC) and the temporoparietal junction (TPJ) that are critical for social perception and empathy.

When a nonhuman object displays human-like attributes, it not only catches our attention but also ignites the neural correlates that are typically reserved for human interactions. A study by Gobbini et al. (2007) indicates that human-like characteristics in robots or cartoon animals induce activation of the fusiform face area (FFA), a brain area commonly linked to face perception and identification.

The interaction between flourishing technology and neural predispositions is becoming an increasingly researched topic in neuroscience. Human-like robots, such as those designed by Ishiguro (2006), challenge our perception and often cause a heightened emotional response and a sense of connection by mimicking human qualities like eye contact or speech patterns, engaging the same circuitry we use for human-to-human interaction.

The aspect of unexpectedness also plays a significant role when we encounter anthropomorphic phenomena. Theory of predictive coding by Friston (2005) posits that the brain is a prediction machine, always trying to minimize the difference between expected and actual stimuli. When predictions fail, as with a speaking car or a problem-solving computer, the resulting surprise increases the likelihood of memory formation and sharing.

These neural mechanisms underpin the allure of anthropomorphic content in the digital world, but how does it translate into application? Marketers and content creators can amplify engagement with their audience by leveraging the natural human inclination to anthropomorphize, thus creating more relatable and shareable content.

Consider the appeal of mascots and brand characters: characters like Tony the Tiger, the Geico Gecko, or the Michelin Man have become cultural icons not just through branding but also by invoking this neural tendency to interact with nonhumans as if they were part of our social circle. Creating a narrative that includes these characters can deepen brand attachment by activating the social brain (Knapp & Corina 2010).

When it comes to advertising that goes viral, infusing products with personality can transform them from mere objects to quirky partners in crime (Nikolinakou & King 2018). It's not just a vacuum cleaner; it's your diligent little helper. This sort of clever personification can tickle the audience's fancy and foster a greater emotional bond to the product. Remember how the internet fell in love with the Mars rovers? *Spirit, Opportunity,* and *Curiosity* became intrepid explorers, not just machines. The practice of giving life to vehicles, appliances, or brands is not a new trend, but there's a renewed sense of charm with advancements in technology. A car that "winks" with its headlights, or a voice-activated assistant that offers witty retorts, capitalizes on this charm by marrying technology with anthropomorphism to make everyday experiences delightful.

It is evident that the brain's wiring for sociability can make a viral superstar out of nearly anything. The next time a video of a puppy appearing to "dance" or a robot assistant telling jokes goes trending, remember it's our own social brains behind the scenes, finding affinity and amusement in the "almost human."

Tapping into our instinct to anthropomorphize can transform marketing efforts from simply informative to irresistibly engaging. By nudging the social areas of the viewers' brains, creators and marketers can make their content spread like wildfire – as every share is a nod to our deeply ingrained love for almost all humans in an increasingly mechanized world.

Use GenAi to design unique and engaging mascots that resonate with the target audience. GenAi can assist in generating character concepts, visual designs, and personality traits that appeal to consumers and evoke positive emotions. Develop mascots or almost human characters that evoke empathy, humor, or nostalgia in advertising campaigns. GenAi can help create characters with relatable traits, mannerisms, and experiences that resonate with consumers on a personal level, fostering emotional connections and increasing the likelihood of virality.

Against the Clock: The Viral Charm of Age Defying Acts

The allure of children displaying mature behaviors and the elderly engaging in youthful activities lies deep within our social cognition mechanisms. Developmental neuroscience suggests that our brains are tuned to recognize and be engaged by typical age-related behaviors. Consequently, when someone defies these expectations, it triggers enhanced neural processing due to the element of surprise.

Research into the theory of mind indicates that humans have an innate ability to attribute mental states to others, which helps in understanding and predicting their behavior (Nguyen et al. 2023). This ability leads us to have certain expectations of individuals based on age-related social norms. Thus, the elderly showing agility or children engaging in complex conversations prompts a reevaluation of these norms, captivating our attention.

From an evolutionary standpoint, observing individuals acting outside of their expected age norms may have served as a crucial cognitive signal for adaptive learning. It could suggest unique skill sets or knowledge worth paying attention to for survival benefit. In the modern context, this translates to a strong social interest and the potential viral spread of such content.

Neuroplasticity, the capacity of the brain to change and adapt, is often associated with the younger brains but is also present in the elderly (Erickson et al. 2013). The unexpected nature of seniors performing extraordinary feats (like marathon running) or toddlers showcasing advanced abilities (like playing a musical instrument) serves as a testament to and reminder of the brain's lifelong adaptability.

Emotion theory asserts that content arousing strong positive emotions, such as humor or awe, is more likely to be shared (Campo et al. 2013). Videos of young children arguing about ethics or old individuals breakdancing offer a juxtaposition that elicits such emotions, enhancing the likelihood of virality.

Transitioning to application, marketers can capitalize on this principle by featuring age-defying individuals in their content strategies. A deliberate tilt could be toward the inspirational factor, which resonates deeply within most audiences and can propel brand messaging within the umbrella of motivation and perseverance.

These contrasts not only entertain – they inspire. They offer a refreshing narrative that aligns with the "age is just a number" philosophy, replete with surprise and a touch of rebellion against societal norms. In the market of attention, these moments stand out, engaging consumers on a level that traditional marketing may not.

In aligning this concept with a brand or product, one might portray the product as an enabler for breaking through the age barrier – be it a tech gadget that an elder can master or a service that empowers a child prodigy. It's about creating undertones of mastery over time, often with a wink to the audience.

Content that is bold enough to push the bounds of time can be intriguing. These viral occurrences are timeless marvels in the digital scroll of time because they may be used as testaments to the potential of people and the limitless nature of ability, humor, and emotion.

Use GenAi to generate creative concepts and scenarios that challenge age stereotypes and expectations. Employ humor and satire to challenge age-related stereotypes and perceptions in advertising campaigns. Use GenAi to develop comedic content that pokes fun at societal expectations and norms surrounding age, prompting laughter and generating buzz among viewers of all ages.

Realms Converged: The Alchemy of Virality in the Melding of Worlds

The human fascination with the blurring boundaries between fantasy and reality is increasingly evident in our digital culture. From the metaverse to immersive gaming, we are captivated by experiences that mix the real with the surreal. Grounded in the science of escapism, studies such as those by Klinger (1990) explain our intrinsic desire to engage with fantasy as a means to explore alternative realities and enhance our sense of well-being.

At the heart of this intersection lies cognitive dissonance, a psychological phenomenon where encountering conflicting beliefs, ideas, or realities creates a state of mental discomfort (Levy et al. 2018). This same dissonance can intrigue and entertain when experienced in a controlled manner, as seen in magical realism in literature or the visual paradox of M. C. Escher's artwork. The success of the Marvel series stems from the violent collisions of real and fantastic worlds. We might argue that Harry Potter also successfully merged the real and the surreal.

The brain's ability to process and reconcile disparate pieces of information involves complex neural networks. The prefrontal cortex, responsible for decision-making and reality testing, interacts with the limbic system, where emotions and memories are formed, to create a coherent narrative from disordered elements. When fantasy elements are seamlessly woven into a real-world setting, our brains work overtime to resolve the inherent contradictions, making such content inherently memorable and shareable.

Technological advances have enabled the creation of lifelike virtual environments that experiment with our perception of reality. Research by Slater & Sanchez-Vives (2016) has shown that virtual reality (VR) can profoundly impact the brain, often leading to a suspension of disbelief and the sensation of having experiences that are almost real. This persuasive illusion can elicit strong emotional responses and create narratives that resonate deeply with the audience (Barnes 2016).

Within this landscape, the concept of liminality plays a crucial role. The liminal space, characterized by the threshold between the known and the unknown, often reveals our fascination with the merging of worlds. The brain enjoys the challenge of filling in gaps in ambiguous narratives, such as Rowling's Platform $\frac{93}{4}$, utilizing imagination and prior knowledge to construct a plausible reality.

In transitioning to a more applied narrative, marketing strategies often play on this intersection of wonder and logic. Consider the commercials that intersperse fantastical elements – unicorns, talking animals, or flying cars – into everyday scenes. Not only do these create an indelible impression, but they also engender a sense of joy and whimsy that enhances the sharing potential of the content.

Virality originates at this nexus, blending the absurd with the mundane. There's something inherently funny about a dragon working an office job or a wizard puzzling over a smartphone. This humor comes from cognitive incongruity, where our brains are tickled by the unexpected merger of contrasting elements. Keeping the content lighthearted, marketers can generate a buzz by juxtaposing the extraordinary with the ordinary.

Take it a step further with application, and we delve into the world of gamification in advertising. Marketers have recognized that by gamifying an experience, weaving narrative game elements into nongame contexts, they can drive engagement through curiosity and the delight of discovery. Encouraging consumers to unlock "magic" in everyday products captures the essence of fantasy-realism that ignites virality.

Melding fantasy with reality in media and marketing is a powerful alchemy that appeals to our neural and narrative desires. It leverages our cognitive biases, fires up our imagination, and activates our social sharing instincts, turning content into digital folklore that transcends the ordinary, one shared experience at a time.

Use GenAi to craft storytelling elements that transport viewers into fantastical worlds while maintaining a connection to real-life experiences and emotions, creating a sense of wonder and intrigue that encourages sharing and engagement.

Generate visually stunning effects and animations that blur the lines between fantasy and reality in advertising campaigns. Use GenAi to create mesmerizing visuals that depict magical landscapes, mythical creatures, or supernatural phenomena, captivating viewers and inspiring them to share the content with others.

Heroes Among Us: The Viral Spark of Everyday Extraordinary

Human excellence, whether showcased through art, talent, athleticism, or acts of kindness, possesses a magnetic allure. At its core, our fascination with extremes can be traced back to the concept of *peak shift*, a term in neural theory where exaggerated versions of a stimulus are preferred over more realistic or moderate ones. This neural bias toward extreme stimuli is a clue to why we are drawn toward exceptional feats.

When discussing talent, the conversation cannot escape the realm of innate abilities and learning. Ericsson's theory of *deliberate practice* posits that it takes around 10,000 hours of focused training to achieve mastery in a field (Ericsson et al. 1993). This notion resounds with our recognition of skill extremes as evidence of not only innate talent but also dedication and perseverance, highlighting the stories that most often go viral.

Kindness, too, is subject to extremes. The neurohormone oxytocin, often linked with social bonding and trust, is released during acts of kindness both in the giver and receiver – and even in onlookers. This biochemical reaction might partly explain the virality of such acts, as kindness resonates profoundly with our biological constructs of community and altruism.

Moving toward application, extreme skill and generosity become more than actions; they are stories that engage our emotions and cognitive grappling with excellence. Memorable content often features acts that challenge perceived limits, thereby stimulating the brain's reward centers, including the mesolimbic pathway, which plays a significant role in motivation and reward-related behavior.

In a marketing context, featuring extraordinary talent or kindness evokes the psychological phenomenon of elevation (Snow et al. 2018). Observing virtuosic performances or profound generosity engenders a feeling of elevation in viewers, fostering a desire to share and emulate those actions. This emotional contagion can transform content into a viral sensation.

Highlighting extremes of athleticism doesn't just show physical prowess; it can tell a story of human potential and resilience. Similarly, showcasing extreme artistry not only reflects skill but can also evoke deep emotional responses that transcend cultural barriers. Both can be used effectively to engage audiences in a narrative that links to the product or brand.

The synergy of extreme talents and marketing follows a simple yet profound pattern: capture attention with the extraordinary, engage emotion with storytelling, and inspire action with relatability. This narrative arc mirrors our intrinsic fascination with extremes and has the potential to catapult brand visibility through virality.

The phenomenon of extremes and their propensity to go viral are documented and known. While science underpins our universal pull toward exceptional feats, it is the human touch in content creation that truly makes virtuosity viral – a formula for success that resonates across digital landscapes.

Craft narratives that highlight the journey, passion, and dedication of extraordinary talent in advertising campaigns. Use GenAi to develop storytelling elements that evoke emotions, inspire awe, and showcase the talent's unique journey and contributions to their respective fields. Provide behind-the-scenes glimpses into the lives and creative processes of extraordinary talent featured in advertising campaigns. Use GenAi to create engaging and authentic content that offers insights into the talent's daily routines, training regimens, and creative inspirations.

The Wild Connection: Beasts Go Viral

In the rich tapestry of the animal kingdom, behaviors such as hunting, mating rituals, and nurturing their young are not merely survival strategies but also fascinating windows into their world. Anthropomorphism suggests that humans attribute human-like traits to animals. This innate tendency is influenced by our evolutionary need to recognize and empathize with living beings, particularly those that display similar social behaviors.

The mirror neuron system, which sparks a neural resonance when observing others, is not exclusive to humans. Although still a topic of research, there's evidence to suggest that these neurons might exist across species. When we watch a lioness care for her cubs or a peacock's elaborate mating dance, we may experience a form of kinship that resonates with our own parental instincts or the complexities of human courtship.

The strategic deployment of visual cues in animal behavior, like the vibrant plumage of a bird during mating season, leverages the human sensitivity to attention-grabbing stimuli. Neuroaesthetics, a field exploring the brain's response to art and beauty, may shed light on why such spectacular displays captivate us and go viral when shared online.

Predatory behaviors in the wild often trigger an adrenaline-fueled response in human observers, harkening back to our primal selves. The neurotransmitter norepinephrine, associated with the fight-or-flight response, floods our system when we witness a gripping chase, causing heightened alertness and emotional arousal prime for sharing.

The caring behaviors of animals juxtapose the predatory ones and tap into our inherent drive toward altruism. The release of oxytocin during such tender moments, observed across various species, including humans, reinforces social bonds. Known as the "love hormone," oxytocin's influence on empathy and attachment suggests why the care of offspring in the animal world tugs at our heartstrings.

Translate these instincts into marketing applications and you get a series of vignettes where products or services align with an animal's vivacious or nurturing traits. A bank might draw on the protectiveness of a wolf pack to demonstrate security, or a dating app could showcase the theatrics of a mating dance to illustrate the excitement of finding a match.

Imagine trying to teach a cat internet security. Its indifferent gaze at the camera as it nonchalantly pushes over a password reminder, scripts the narrative for online privacy in a humorous twist that underscores both the importance of protection and the casual neglect we sometimes show it.

The virality of animal content is a seamless blend of innate neuroscientific responses and the whimsical joy of witnessing the untamed mirrored in our digital lives. Sharing these moments serves as a virtual extension of our instinctual fascination with traits that, while distinctly wild, feel strikingly human.

Use GenAi to create visually stunning imagery and videos featuring animals that capture the attention of viewers. Utilize high-quality photography and videography techniques to showcase the beauty, personality, and behavior of animals in a captivating and engaging manner. Incorporate humor and entertainment into advertising campaigns featuring animals to engage and entertain viewers. Use GenAi to develop comedic content, memes, or viral challenges that showcase the playful and endearing aspects of animals, encouraging sharing and engagement across social media platforms.

Stars in Line: The Neuroscience of Celebrity Daily Events and Mishaps Going Viral

The virality of celebrities caught in the lens of the ordinary owes much to a psychological phenomenon known as the *pratfall effect*. Coined by psychologist Elliot Aronson in 1966, the pratfall effect suggests that competent individuals become more likable when they make a mistake. It humanizes them, making them more relatable to the general public.

When a recognized figure engages in ordinary, relatable activities or even commits a faux pas, it triggers a sense of familiarity and shared experience within observers. This cognitive reassessment – viewing the untouchable as touchingly real – is underlined by what psychologists call the paradox of fame and intimacy.

Comedians Getting Coffee is not just a show, it is just the extraordinary doing the ordinary – just brilliantly original, and neuroscience aligned.

Neuroscientifically, we process information about known individuals through a cortical network that also processes self-relevant thoughts. Watching a celebrity in a relatable situation activates regions in the medial prefrontal cortex, akin to processing information about ourselves or our close acquaintances.

The concept of embarrassment and empathy also plays a crucial role. Witnessing a celebrity's amusing blunder can cause empathic embarrassment, which involves the brain's affective circuits, such as the anterior insula and anterior cingulate cortex. This shared discomfort further strengthens the social bond between fans and celebrities.

Celebrity-driven campaigns, like the ALS Ice Bucket Challenge, leveraged the social aspect of neuroscience. The dopamine-mediated reward system, including areas like the striatum and ventromedial prefrontal cortex (VMPFC), is triggered in acts of charity, enhancing the psychological payoff for sharing content.

Marketers can conjure virality by incorporating ordinary moments into celebrity endorsements (Ganisasmara & Mani 2020). Picture an advertisement where a pop star gets locked out of their car – cue the brand's roadside assistance saving the day, while wittily jabbing at the star's mortality. Here, the ordinary promotes the product, and humor entwines with the solution.

Such engaging scenarios often embody the "celebrities – they're just like us" maxim. By pairing well-known personas with everyday snafus and the products that resolve them, a relatable narrative is spun. Wrapping up the spectacle, it's the juxtaposition of high celebrity status with everyday slip-ups that fuels the virality engine. These moments resonate because they echo the science of shared human experience – revealing the stars as more than just distant constellations in our personal skies but as companions in the sometimes clumsy dance of life.

Use GenAi to digitally alter existing footage or images of celebrities in a way that adds humorous or entertaining elements without causing embarrassment. This could involve adding playful animations, sound effects, or visual effects to create a lighthearted and engaging experience for viewers.

Collaborate with celebrities to create scripted comedy sketches or advertisements that showcase their comedic talents in a controlled and respectful manner. Use GenAi to assist in scriptwriting and storyboarding, ensuring that the content is humorous and entertaining without crossing any boundaries or causing embarrassment.

Mimicry Unmasked: Breaking Down the Virality of Parody

Parody, an imitation with a comedic twist, is a cultural phenomenon heavily interwoven with our cognitive processes. It plays on familiar templates – gestures, expressions, voices – to create humor through exaggeration and contextual incongruity. This begins with the basic human ability to recognize and interpret faces, a process facilitated by the fusiform gyrus, known as the brain's face area.

Beyond basic recognition, our mirror neuron system is engaged when we perceive actions and expressions, even in a parodied context. Studies demonstrate that these neurons respond not just to actions but also to the intention behind them, providing the scaffolding for understanding parody's exaggerated mimicry.

The phonological loop in our working memory, a component of the Baddeley and Hitch model, plays a crucial role when it comes to the auditory side of parody – processing and retaining the rhythm, pitch, and cadence of voices. Such auditory cues are essential in establishing a character's distinct persona that is ripe for humorous imitation.

Social context profoundly shapes the efficacy of parody, with shared knowledge providing the substrate for humor. Through theory of mind, an aspect of our cognition that allows us to infer others' thoughts and intentions, we appreciate the nuances of satire and sarcasm, often hallmarks of effective parody.

Diving into the realm of mainstream application and entertainment, we can draw a line connecting the science of parody to the viral skits we see on platforms like *Saturday Night Live*. A politician's quirks transformed into hyperbolic sketch comedy, or a pop icon's signature moves taken to outlandish levels, reflect the underlying neural mechanisms at play in humorous recognition.

The appeal of parody also lies in its nature of being a creative twist on reality. Just think of Weird Al Yankovic's legendary parody songs – by matching familiar tunes with new, amusing lyrics, he mastered turning the ordinary into an absurd delight, capitalizing on our love for the unexpected.

Marketers can leverage parody's electric charge by creating ad campaigns that poke fun at their own industry. For instance, commercials that overly dramatize a product's effect with exaggerated satisfaction – the infamous "Got Milk?" ads come to mind – capitalize on this humor.

Parody's magnetic pull in our digital age is as evident as its roots in complex cognitive processes. By imitating, exaggerating, and subverting the known, parodies resonate with our neural circuits, incite communal laughter, and often become the emblems of online virality.

Use GenAi to generate content featuring celebrity impersonations that are entertaining and engaging for viewers. This could involve creating videos or images of celebrities mimicking other famous figures or characters, adding a humorous and relatable element to the advertising campaign.

Create viral challenges or trends that encourage consumers to mimic specific actions or behaviors featured in the advertising campaign. Use GenAi to develop interactive elements, such as augmented reality filters or TikTok-style challenges, that prompt users to participate and share their own interpretations of the content.

Tap into meme culture by generating content that mimics popular memes or internet trends. Use GenAi to create memes or GIFs that playfully reference current events, pop culture phenomena, or viral sensations, making the advertising campaign relatable and shareable among online communities.

Striking Chords: Uncommon Unconventional Music

Music's capacity to resonate with and captivate audiences is steeped in its neural impact. The auditory cortex is crucial for processing the complex layers of sound in music, but when we encounter a universally recognized tune like Beethoven's "Moonlight Sonata" played on a ukulele, it engages a wider neural network. This network involves not only auditory processing but also memory and emotional centers, reflecting the preservation and emotion bound within the familiar melody.

The unexpected instrumentation introduces a novelty factor, activating the brain's reward circuits. The striatum and the orbitofrontal cortex, key components involved in recognizing and processing novelty, light up upon encountering a novel stimulus, such as a classical piece performed on an unconventional instrument.

Further effect of music on the brain comes from its structure and rhythm, often reinforcing motor actions. When rhythm is executed in an unconventional way, such as the intricate picking of a ukulele string to replicate a piano sonata, it triggers activity in the basal ganglia, which contributes to the coordination of movement and the pleasurable response often associated with toe-tapping or following the beat.

Musical expertise, or virtuosity, relies on intense practice that remodels the brain. Structural changes in motor and auditory areas, as shown by studies on musicians, indicate that musical training enhances brain plasticity (Moreno & Bidelman 2014). When a musician transitions a piece intended for one instrument to another, it showcases an extreme level of skill and neural adaptability.

We are also predisposed to appreciate skillful expressive deviations in music, such as timing and dynamics (Karpasitis et al. 2018). When a ukulele player adds his or her own embellishments to a classic track, the brain's insula and limbic structures become activated, aligning the emotional expression in music with internal emotional states.

Like the punchline of a joke hitting home after a winding setup, the punchy strums of a ukulele producing sublime classical sounds prompt a delightful realization and a rush of endorphins. This blend of familiarity and surprise often lands videos of such performances in the viral vortex of social sharing.

Virality of music comes through extreme talent, unexpected musical settings, unexpected instrumentation, the blending of human and animal, the blending of instruments and noise, and the cognitive dissonance of instruments and genres. They create shareable and talkable moments.

Use GenAi to experiment with unconventional music genres, styles, and compositions that defy traditional expectations. Explore avant-garde, experimental, or niche genres to create music that stands out and captivates the audience's attention with its novelty and uniqueness. Blend disparate musical elements and influences using GenAi to create hybrid compositions that defy categorization. Mix genres, cultural styles, and sonic textures to produce music that transcends boundaries and offers a fresh and unexpected listening experience for consumers.

Launch viral challenges or trends that encourage consumers to engage with and share unusual music creations. Use GenAi to develop catchy hooks, melodies, or rhythms that inspire audience participation and sharing across social media platforms, driving organic reach and engagement for the advertising campaign.

Voyeurs within Us: The Viral Science of Seeing the Unseen

The allure of voyeurism triggers the brain's natural curiosity and information-seeking behavior. Witnessing the previously concealed evokes a sense of discovery analogous to primitive foraging behaviors. This arousal is marked by activity in the nucleus accumbens, part of the brain's reward system, signaling the pleasurable payoff when uncovering hidden information.

The act of unveiling the secret also invokes a psychological phenomenon known as the information-gap theory. Curiosity arises when there's a gap between what we know and what we want to know. The driving force to close this gap results in focused attention and engaging cognitive processes, facilitated by the dopaminergic system, which is stimulated by novelty and the unexpected.

Our perception systems, both visual and auditory, are fine-tuned to detect changes in our environment. A hidden camera revealing private actions intersects with the element of surprise – a fundamental characteristic that enhances memory encoding via activation of the amygdala.

Given the prohibition compound aspect, the brain's reaction to forbidden content taps into social norms and moral reasoning. Areas such as the right temporoparietal junction (rTPJ) and the medial prefrontal cortex (mPFC) are especially implicated in making judgments about taboo or socially unacceptable behavior, heightening our attention to such content.

Like a first-person narrative, voyeuristic viral content creates an immersive experience, fostering a sense of empathy and connection. When we privately view private actions, our mirror neuron system fires as if we were experiencing the events ourselves, allowing us to connect on a deeper emotional level, even when the on-screen actions are humorous or lighthearted.

As we move from the psychological underpinnings to social phenomena, think of the infamous show *Candid Camera*. Viewers are drawn to these humorous invasions of privacy because they expose the universally human reactions to oddball scenarios – reactions that the insula aids in appreciating as it processes complex social emotions.

The first-person perspective – often employed in video games and virtual reality – deepens the viewer's engagement. When a marketing campaign applies similar immersive strategies, hidden camera style, it creates a direct experiential link. Imagine an ad where consumers feel they're sneaking into a secret conclave of chefs discovering the new recipe for a popular snack, blending the thrill of discovery with an enticing product reveal.

Unbox and Unravel

The tractive lumina saccharum. That's Latin for "light attracts moths," and in the context of human curiosity, the sudden exposure of the hidden lures us in much the same way. Just as moths are captivated by a flame, the amygdala and the hippocampus are involved in our attraction to novel stimuli, meaning that the unveiling of previously hidden content is primed to seize our attention and memory.

Consider, for example, the fascination with popular unboxing videos. The neural engagement in such content is mirrored by the frontotemporal regions, which process anticipation and build up suspense as we await the reveal. Furthermore, the dopaminergic systems that interlace these areas equate the anticipation to a cognitive crescendo, fueling the desire to reach the climactic unveiling.

The sparks of virality ignite when the unseen are surprisingly showcased. It's a neural stimulation where the curtain drops, the invisible becomes visible, and the private goes public, all to the backdrop of cognitive fireworks that pleases and binds audiences in shared humor and amazement.

Combined with modern digital tactics, this understanding ascends to the level of art – craftily using the architecture of curiosity and the science of exposure to entertain and connect, imprinting the revealed secrets on our collective consciousness.

Develop interactive unboxing videos that allow viewers to engage with the content and explore products in a dynamic and immersive way. Use GenAi to generate interactive elements, such as clickable hotspots, 360-degree views, or augmented reality overlays, that enhance the viewing experience and encourage sharing and engagement across social media platforms.

Incorporate surprise and delight elements into the unboxing experience to create moments of excitement and joy for consumers. Use GenAi to generate unexpected surprises, hidden messages, or interactive elements that delight consumers as they unbox their purchases, leaving a lasting impression and driving word-of-mouth promotion.

Amazing and Incredible Me – Ego Sublime

Content that boosts self-worth or highlights personal branding often has a high propensity to go viral. At the core of this phenomenon is the concept of self-affirmation in social cognitive theory, which posits that individuals are driven to protect the integrity of the self. When online material compliments the audience's self-image or reflects their values, it resonates on a personal level, activating the ventromedial prefrontal cortex (vmPFC), a brain region related to self-referential thinking and valuation.

Dovetailing with the trend toward personal branding, the element of social identity theory comes into play. This psychological framework explains how individuals strive to enhance their self-esteem by maintaining and promoting their social group affiliations. Content that aligns with the viewer's social or aspirational group, be it through common values or collective accolades, organically compels sharing, as individuals seek to manage the social self.

The science of emotional contagion, the process by which emotions are transferred from one person to another, underscores virality tied to emotional extremes. Social media platforms have turned into hotbeds for rapid emotional exchange, where exposure to affect-laden content can incite similar emotional responses in the viewer's brain, particularly within the insula and the cingulate cortex, regions implicated in emotional awareness and empathy.

From a neuroAi perspective, content that achieves virality by enhancing self-conception is meticulously designed to provoke introspection and self-evaluation. The encoding and retrieval of related information involves the activation of the hippocampus within the medial temporal lobe, an area frequently linked to the personal relevance and context of experiences.

Extending into the realm of internet phenomena, personal branding through digital content is ever evolving and is bolstered by the neuronal correlates of reward anticipation. The anticipation of social and self-reward elicits activity in the brain's vast dopaminergic systems, wherein lies the impetus for pursuing goal-directed behavior and the propagation of self-centric content.

Consider the joy that courses through someone when they post a selfie and watch the "likes" flood in (Hartmann et al. 2021). This surge parallels a bump in dopamine release, akin to unwrapping gifts on a birthday. It's the brain's way of patting oneself on the back, and that's the simplest kind of magic – a neurological high-five that we're ever eager to share.

Marketers can promote content that champions individual triumphs or group successes – be it through inspiring testimonials or humorous skits about workplace victories. This path not only increases brand awareness but more importantly, makes the consumer the hero of their own story, fostering brand loyalty.

Viral marketing that elevates personal branding and thrives on emotional extremes capitalizes on deeply ingrained neurocognitive patterns. It's the entertaining blend of self-endorsement with emotional resonance that transmutes ordinary content into a viral goldmine, reinforcing the modern-day dictum: Every individual is a brand, every emotion is a story, and every share is an affirmation.

Use GenAi to generate personalized affirmations and positive messages tailored to individual consumers. Develop algorithms that analyze user data and preferences to create uplifting and empowering content that resonates with each consumer's unique identity and values.

Extremes of Negative Emotion – Embracing the Dark Side

The propagation of content infused with intense negative emotions taps into a primal aspect of human neuroscience. The emotion of anger, for instance, correlates with increased activity in the left prefrontal cortex, which is associated with approach and confrontational behaviors – responses evolutionarily designed for survival. This neural activation can lead to anger-infused content becoming highly shareable as it resonates with our inherent fight response.

Experiences of pain, both physical and emotional, spark noteworthy activity in the anterior cingulate cortex (ACC) and the insula, brain regions involved in discomfort and empathy. Observers vicariously experience the pain displayed in viral content, invoking a potent empathic response that can motivate the sharing of content to seek consolation or communal understanding.

Fear has its own niche in the viral world due to its activation of the amygdala, the brain's alarm system for processing threats. Fear-based content often goes viral because it stimulates the amygdala's fast pathway, bypassing rational thought and leading to instinctual sharing, presumably as an evolutionary mechanism to quickly inform others of potential danger.

Disgust is processed by the basal ganglia, along with the insula, and has been linked to a deeply rooted avoidance behavior. It elicits a strong, almost visceral reaction to stimuli considered offensive or contaminative, which in the realm of viral content can paradoxically become attractive due to the intense emotional response it invokes.

Sadness connects deeply with listeners through the release of the hormone prolactin, which plays a role in consoling and soothing behaviors (Huron 2011). Sharing content that elicits sadness can thus be a cathartic experience, offering a sense of emotional relief and a subconscious quest for empathy and connection, explaining its viral tendency.

The scientific underpinnings of negativity's virality alternate with our modern compulsion to share. Imagine the potent cocktail of witnessing a public figure's outrage spilling over into a heated tirade – mismatched with the harmless sulking of a disappointed bulldog. It's a tragi-comic symphony that our brains are all too eager to conduct across the social media spectrum.

The meme culture, thriving on hyperbolic portrayals of life's downtrodden moments, often brings laughter by simply exaggerating to the point of absurdity – the angrier the internet cat, the louder our collective chuckle.

From a marketing angle, cleverly designed content that capitalizes on these powerful emotions can become especially sticky. The iconic "Fearless Girl" statue, facing Wall Street's Charging Bull, taps into a shared sentiment of brave defiance. This juxtaposition playfully tugs at our fight response, turning a poignant symbol into a viral sensation.

Create narratives that highlight in an extreme way a relatable problem or pain point faced by the target audience, followed by the presentation of a solution provided by the advertised product or service. GenAi can assist in crafting emotionally resonant storytelling that effectively communicates the problem and the proposed solution.

Extremes of Positive Emotion: Embracing the Light Within

The experience of joy engages several complex networks in the brain, including the pursuit of rewards and the processing of positive emotion. One critical area implicated in the sensation of joy is the orbitofrontal cortex (OFC), which becomes active during rewarding experiences and is associated with the evaluation of pleasure. This activation may explain why joyous content has a high likelihood of being shared; it simply feels good to spread happiness.

Surprise has a powerful hold on our attention and memory. Thanks to the activation of the amygdala when surprised, our emotional response is heightened and makes such content more shareable. Pleasant surprises, such as unexpected positive outcomes or twists in a storyline, create a "eureka" effect of delight that heavily capitalizes on social media dynamics.

Love, another potent positive emotion, enraptures and captivates. When exposed to romantic or love-filled content, the brain's ventral tegmental area (VTA), a component of its reward system, is triggered, releasing dopamine. Dopaminergic activation compels users to engage with and propagate content that echoes their emotional state.

Desire, often oriented toward attractive stimuli, activates the striatum, where we process rewarding experiences. The visual allure of aesthetically pleasing imagery or filmography resonates deeply on a neural level, encouraging the sharing of content that aligns with the viewer's aspirational desires.

Utilize GenAi to generate extremes of positive emotion. Develop interactive comedy experiences that engage consumers in humorous storytelling or gameplay. Use GenAi to create interactive elements, such as chatbots, quizzes, or augmented reality filters, that allow consumers to participate in comedic scenarios and share their experiences with friends and followers.

In this chapter we have covered the core elements that make content go viral. Depending on the brand, the culture, and the product, GenAi algorithms can and must be guided to generate viral content.

CHAPTER 20

Imagine Our Future

O ur vision for the future is both optimistic and bold. We view GenAi powered by neuroscience as the true liberator. We all live lives that are shackled – our creativity confined to narrow domains of expertise – while we all experience a longing to explore so many more domains in which we remain untrained. GenAi unleashes our creativity and becomes our guide into territories we know nothing about.

Imagine an enterprise where a young woman sitting in Mumbai with the help of neuroscience-powered GenAi can imagine the needs of a young woman in New York City and design products and services for her.

Imagine an organization where every employee uses GenAi to get trained in every category and product line.

Imagine a CPG company spending big money on the 20% of its brands that generate 80% of its profits but using GenAi to support the 80% of its brands that suffer from a lack of investment.

Imagine a company that ends the purchase of music but generates its own music for its ads using musicGPT.

Imagine a company that uses the engines of desire to evaluate every image, every copy of its billions of content on digital sites, using desireGPT, and then scoring it, and rewriting it to evoke desire.

Imagine an enterprise where customer understanding, product innovation, and packaging design even as inspiration are all done in a few minutes using trendGPT.

Imagine a company where every product innovation automatically generates an associated provisional patent that protects the intellectual property – digital factories that generate IP and patents.

Imagine an enterprise where organizational memory and knowledge is coded as rules, embeddings, and transformers – expertise that does not walk out the door.

Imagine a world where data lakes and warehouses transform from housing data to becoming transformer models that become tools for daily use – where data become tools and products that generate revenue and competitive advantage.

Imagine an enterprise where the fragrance and flavor design is democratized and even co-created with the consumer using fragranceGPT and flavorGPT.

Imagine a world where every company is its own creative agency and generates all the necessary short form, long form video and audio and point of sale materials with a few keystrokes of adGPT.

Imagine a world where branding powered by brandGPT generates brand love and loyalty.

Reimagine packages optimized algorithmically, created by packageGPT, that arrest attention in the aisle and evoke purchase intent.

Imagine neuroscience-powered GenAi enabling us to realize the inner artist and creative genius in all of us, without the drudgery of discipline, experience, or practice.

Imagine customer experience dramatically enhanced at every touchpoint through a blend of neuroscience and GenAi.

Imagine newer patterns, paradigms and methods discovered by GenAi in disciplines as rigorous as mathematics, physics, and biology – metaphor-powered discovery of scientific paradigms

Imagine John Lennon using GenAi to write the lyrics to "Imagine" – which is exactly what we did.

(Verse 1)
Imagine there's no borders
It isn't hard to do
No lines between mind and machine,
And no prejudices, too.
Imagine all the people
Living life with A.I. . .
(Chorus)
You may say I'm a dreamer,
But I'm not the only one.
I hope someday you'll join us,
In a world where mind and code unite as one.
(Verse 2)
Imagine there's no limit,
To what both could achieve
No need for greed or hunger,
A world of open knowledge, believe.
Imagine all the people,
Sharing wisdom in a hive. . .
(Chorus)
You may say I'm a dreamer,
But I'm not the only one
I hope someday you'll join us,
And the world of GenAi will thrive.
(Bridge)
Imagine understanding,
The deepest depths of mind,
With neuroAi in our daily lives,
Oh, the truths we could find.
Imagine all the people
Unleashing genius, all the time. . .

NeuroAi is not about replacing people or jobs. It takes us to a brave new world where our jobs suit our passions, no job is beyond us, every stroke becomes a masterstroke, and no lesson learned is ever forgotten.

We welcome you, dear reader, to this brave new world, and wish you a wonderful journey of discovery of yourself on this road.

References

Aaker, David. *Creating Signature Stories: Strategic Messaging that Persuades, Energizes and Inspires*. Morgan James Publishing, 2018.

Abdolmohamad Sagha, M., Seyyedamiri, N., Foroudi, P., & Akbari, M. (2022). The one thing you need to change is emotions: The effect of multi-sensory marketing on consumer behavior. *Sustainability* 14(4): 2334.

Acerbi, A., & Stubbersfield, J. M. (2023). Large language models show human-like content biases in transmission chain experiments. *Proceedings of the National Academy of Sciences* 120(44): e2313790120.

Achiam, J., Adler, S., Agarwal, S., Ahmad, L., Akkaya, I., Aleman, F. L., ... & McGrew, B. (2023). Gpt-4 technical report. *arXiv preprint arXiv:2303.08774*.

Adolphs, R., Tranel, D., Damasio, H., & Damasio, A. R. (1995). Fear and the human amygdala. *Journal of Neuroscience* 15(9): 5879–5891.

Adriatico, J. M., Cruz, A., Tiong, R. C., & Racho-Sabugo, C. R. (2022). An analysis on the impact of choice overload to consumer decision paralysis. *Journal of Economics, Finance and Accounting Studies* 4(1): 55–75.

Aggleton, J. P., Waskett, L. (1999). The ability of odours to serve as state-dependent cues for real-world memories: can Viking smells aid the recall of Viking experiences? *British Journal of Psychology* Feb;90 (Pt 1):1–7. doi: 10.1348/000712699161170. PMID: 10085542.

Agrawal, V. (2022). Biomarketing: Human body as marketing engine. doi: 10.2139/ssrn.4079030.

Ahn, Y. Y., Ahnert, S. E., Bagrow, J. P., & Barabási, A. L. (2011). Flavor network and the principles of food pairing. *Scientific Reports* 1(1): 196.

Alexander, B. (2017). *The New Digital Storytelling: Creating Narratives with New Media*. Revised and updated edition. Bloomsbury Publishing USA.

Algoe, S. B., Kurtz, L. E., & Grewen, K. (2017). Oxytocin and social bonds: The role of oxytocin in perceptions of romantic partners' bonding behavior. *Psychological Science* 28(12): 1763–1772.

Allen, C. T., Fournier, S., & Miller, F. (2018). Brands and their meaning makers. In *Handbook of Consumer Psychology* (pp. 773–814). Routledge.

Alsharif, A. H., Salleh, N. Z. M., Alrawad, M., & Lutfi, A. (2023). Exploring global trends and future directions in advertising research: A focus on consumer behavior. *Current Psychology*, 1–24.

Alter, A. L., & Oppenheimer, D. M. (2008). Easy on the mind, easy on the wallet: The roles of familiarity and processing fluency in valuation judgments. *Psychonomic Bulletin & Review* 15(5): 985–990.

Anderson, A. K. (2023). Ask the Storyteller: Calling the Archetype to Center Stage. Doctoral dissertation, Pacifica Graduate Institute.

Andreano, J. M., & Cahill, L. (2009). Sex influences on the neurobiology of learning and memory. *Learning & Memory* 16(4): 248–266.

Ariely D, Berns GS. (2010). Neuromarketing: the hope and hype of neuroimaging in business. *Nat Rev Neurosci*. Apr 11(4): 284–292. doi:10.1038/nrn2795. Epub 2010 Mar 3. PMID: 20197790; PMCID: PMC 2875927.

Arnett, J. J. (2007). Emerging adulthood: What is it, and what is it good for? *Child Development Perspectives* 1(2): 68–73.

Aronson, E., Willerman, B., & Floyd, J. (1966). The effect of a pratfall on increasing interpersonal attractiveness. *Psychonomic Science* 4(6): 227–228.

Asamoah, J., Galpin, A., & Heinze, A. (2016). What is video virality? An introduction to virality metrics. *UK Academy for Information Systems Annual Conference*.

Ayres, A. J. (1972). Improving Academic Scores through Sensory Integration. *Journal of Learning Disabilities* 5(6): 338–343. **https://doi.org/10.1177/002221947200500605**

Baddeley, A., Thomson, N., & Buchanan, M. (1975). Word length and the structure of short-term memory. *Journal of Verbal Learning and Verbal Behavior* 14(6): 575–589.

Balconi, M., Venturella, I., Sebastiani, R., & Angioletti, L. (2021). Touching to feel: brain activity during in-store consumer experience. *Frontiers in Psychology* 12: 653011.

Bandettini PA. (2009). What's new in neuroimaging methods? *Ann N Y Acad Sci.* Mar; 1156: 260–293. doi: 10.1111/j.1749-6632.2009.04420.x. PMID: 19338512; PMCID: PMC2716071.

Bandura, A. (2001). Social cognitive theory: An agentic perspective. *Annual Review of Psychology* 52(1): 1–26.

Banskota, S., Ghia, J. E., & Khan, W. I. (2019). Serotonin in the gut: Blessing or a curse. *Biochimie* 161: 56–64.

Bargh, J. (2017). *Before You Know It: The Unconscious Reasons We Do What We Do.* Simon and Schuster.

Baron-Cohen, S., Knickmeyer, R. C., & Belmonte, M. K. (2005). Sex differences in the brain: implications for explaining autism. *Science* 310(5749): 819–823.

Barnes, S. (2016). Understanding virtual reality in marketing: Nature, implications and potential. *Implications and Potential* (November 3, 2016).

Batat, W. (2019). *Experiential Marketing: Consumer Behavior, Customer Experience and the 7Es.* Routledge.

Batey, M. (2015). *Brand Meaning: Meaning, Myth and Mystique in Today's Brands.* Routledge.

Baumeister, R. F., & Leary, M. R. (2017). The need to belong: Desire for interpersonal attachments as a fundamental human motivation. *Interpersonal Development,* 57–89.

Beard, E., Henninger, N. M., & Venkatraman, V. (2024). Making ads stick: Role of metaphors in improving advertising memory. *Journal of Advertising* 53(1): 86–103.

Bechara, A., & Damasio, A. R. (2005). The somatic marker hypothesis: A neural theory of economic decision. *Games and Economic Behavior* 52(2): 336–372.

Bekhbat, M., & Neigh, G. N. (2018). Sex differences in the neuro-immune consequences of stress: Focus on depression and anxiety. *Brain, Behavior, and Immunity* 67: 1–12.

Belfi, A. M., & Jakubowski, K. (2021). Music and Autobiographical Memory. *Music & Science* 4. **https://doi.org/10.1177/20592043211047123**

Belfi, A. M., Karlan, B., & Tranel, D. (2016). Music evokes vivid autobiographical memories. *Memory* 24(7): 979–989.

Bender, E. M., Gebru, T., McMillan-Major, A., & Shmitchell, S. (2021, March). On the dangers of stochastic parrots: Can language models be too big. In *Proceedings of the 2021 ACM Conference on Fairness, Accountability, and Transparency* (pp. 610–623).

Bentley, P. R., Fisher, J. C., Dallimer, M., Fish, R. D., Austen, G. E., Irvine, K. N., & Davies, Z. G. (2023). Nature, smells, and human wellbeing. *Ambio* 52(1): 1–14.

Bercea, M. D. (2012, August). Anatomy of methodologies for measuring consumer behavior in neuromarketing research. In *Proceedings of the Lupcon Center for Business Research (LCBR) European Marketing Conference.* Ebermannstadt, Germany.

Bhatia, T. K. (2019). Emotions and language in advertising. *World Englishes* 38(3): 435–449.

Bjertrup, A. J., Friis, N. K., & Miskowiak, K. W. (2019). The maternal brain: neural responses to infants in mothers with and without mood disorder. *Neuroscience & Biobehavioral Reviews* 107: 196–207.

Blakemore, S. J., Burnett, S., & Dahl, R. E. (2010). The role of puberty in the developing adolescent brain. *Human Brain Mapping* 31(6): 926–933.

Braem, S., & Egner, T. (2018). Getting a grip on cognitive flexibility. *Current directions in psychological Science* 27(6): 470–476.

Brielmann, A. A., Buras, N. H., Salingaros, N. A., & Taylor, R. P. (2022). What happens in your brain when you walk down the street? Implications of architectural proportions, biophilia, and fractal geometry for urban science. *Urban Science* 6(1): 3

Brown, T., Mann, B., Ryder, N., Subbiah, M., Kaplan, J. D., Dhariwal, P., ... & Amodei, D. (2020). Language models are few-shot learners. *Advances in neural information processing systems* 33: 1877–1901.

Bubeck, S., Chandrasekaran, V., Eldan, R., Gehrke, J., Horvitz, E., Kamar, E., . . . & Zhang, Y. (2023). Sparks of artificial general intelligence: Early experiments with gpt-4. *arXiv preprint arXiv:2303.12712.*

Buçinca Z, Malaya, M. B., & Gajos, K. Z. 2021. To Trust or to Think: Cognitive Forcing Functions Can Reduce Overreliance on AI in AI-assisted Decision-making. Proc. ACM Hum.-Comput. Interact. 5, CSCW1, Article 188 (April 2021): 21 pages. **https://doi.org/10.1145/3449287**

Buzsáki, G. (2006). *Rhythms of the Brain.* New York, Oxford Academic. **https://doi.org/10.1093/acprof:oso/9780195301069.001.0001**, accessed 14 Mar. 2024.

Bzdok, D., Langner, R., Schilbach, L., Engemann, D. A., Laird, A. R., Fox, P. T., & Eickhoff, S. B. (2013). Segregation of the human medial prefrontal cortex in social cognition. *Frontiers in* Human *Neuroscience* 7: 232.

Cahill, L. (2006). Why sex matters for neuroscience. *Nature Reviews Neuroscience* 7(6): 477–484.

Camerer, C., et al. (2004). Neuroeconomics: How neuroscience can inform economics. *Journal of Economic Literature* 43(1): 9–64.

Campo, S., Askelson, N. M., Spies, E. L., Boxer, C., Scharp, K. M., & Losch, M. E. (2013). "Wow, that was funny" the value of exposure and humor in fostering campaign message sharing. *Social Marketing Quarterly* 19(2): 84–96.

Cañas, J., Quesada, J., Antoli, A., & Fajardo, I. (2003). Cognitive flexibility and adaptability to environmental changes in dynamic complex problem-solving tasks. *Ergonomics* 46(5): 482–501.

Carr, A. (2015). *The Handbook of Child and Adolescent Clinical Psychology: A Contextual Approach.* Routledge.

Carstensen, L. L. (2021). Socioemotional selectivity theory: The role of perceived endings in human motivation. *The Gerontologist* 61(8): 1188–1196.

Casado-Aranda, L. A., Van der Laan, L. N., & Sánchez-Fernández, J. (2018). Neural correlates of gender congruence in audiovisual commercials for gender-targeted products: An fMRI study. *Human Brain Mapping* 39(11): 4360–4372.

Casey, B. J., Jones, R. M., & Hare, T. A. (2008). The adolescent brain. *Annals of the New York Academy of Sciences* 1124(1): 111–126.

Castelvecchi, D. (2023). How will AI change mathematics? *Nature* 615: 15–16.

Cavalcante Siebert, L., Lupetti, M.L., Aizenberg, E. *et al.* (2023). Meaningful human control: actionable properties for AI system development. *AI Ethics* 3: 241–255 **https://doi.org/10.1007/s43681-022-00167-3**

Chandra, A., Tünnermann, L., Löfstedt, T., & Gratz, R. (2023). Transformer-based deep learning for predicting protein properties in the life sciences. *Elife* 12: e82819.

Chapman, H. A., & Anderson, A. K. (2012). Understanding disgust. *Annals of the New York Academy of Sciences* 1251(1): 62–76.

Charles, S. T., Mogle, J., Leger, K. A., & Almeida, D. M. (2019). Age and the factor structure of emotional experience in adulthood. *The Journals of Gerontology: Series B* 74(3): 419–429.

Chatterjee, A., & Vartanian, O. (2014). Neuroaesthetics. *Trends in Cognitive Sciences* 18(7): 370–375.

Chee, Q. W., & Goh, W. D. (2018). What explains the von Restorff effect? Contrasting distinctive processing and retrieval cue efficacy. *Journal of Memory and Language* 99: 49–61.

Chu, S., & Downes, J. J. (2000). Odour-evoked autobiographical memories: Psychological investigations of Proustian phenomena. *Chemical Senses* 25(1): 111–116.

Ciorciari, J., Pfeifer, J., & Gountas, J. (2019). An EEG study on emotional intelligence and advertising message effectiveness. *Behavioral Sciences* 9(8): 88.

Cirillo, J. (2004). Communication by unvoiced speech: The role of whispering. *Anais da Academia Brasileira de Ciências* 76: 413–423.

Commodari, E., & Guarnera, M. (2008). Attention and aging. *Aging Clinical and Experimental Research* 20: 578–584.

Cooper, R. A., Kensinger, E. A., & Ritchey, M. (2019). Memories fade: The relationship between memory vividness and remembered visual salience. *Psychological Science* 30(5): 657–668.

Coray, R., & Quednow, B. B. (2022). The role of serotonin in declarative memory: A systematic review of animal and human research. *Neuroscience & Biobehavioral Reviews* 139: 104729.

Cowan, N. (1988). Evolving conceptions of memory storage, selective attention, and their mutual constraints within the human information-processing system. *Psychological Bulletin* 104(2): 163–191.

Craig, A. D. (2009). How do you feel—now? The anterior insula and human awareness. *Nature Reviews Neuroscience* 10(1): 59–70.

Creswell, J. D., Bursley, J. K., Satpute, A. B. (2013). Neural reactivation links unconscious thought to decision-making performance. *Social Cognitive and Affective Neurosci*ence Dec; 8(8): 863–869. doi: 10.1093/scan/nst004. Epub 2013 Jan 12.

Cross, S. E., & Madson, L. (1997). Models of the self: Self-construals and gender. *Psychological Bulletin* 122(1): 5.

Cservenka, A., Herting, M. M., Seghete, K. L. M., Hudson, K. A., & Nagel, B. J. (2013). High and low sensation seeking adolescents show distinct patterns of brain activity during reward processing. *Neuroimage* 66: 184–193.

Damasio, Antonio. *Feeling & Knowing: Making Minds Conscious.* Pantheon, 2021.

Das, M., Balaji, M. S., Paul, S., & Saha, V. (2023). Being unconventional: The impact of unconventional packaging messages on impulsive purchases. *Psychology & Marketing* 40(10): 1913–1932.

Daugherty, T., Hoffman, E., Kennedy, K. and Nolan, M. (2018). Measuring consumer neural activation to differentiate cognitive processing of advertising: Revisiting Krugman. *European Journal of Marketing* 52(1/2): 182–198. **https://doi.org/10.1108/EJM-10-2017-0657**

Davranche, K., Nazarian, B., Vidal, F., & Coull, J. (2011). Orienting attention in time activates left intraparietal sulcus for both perceptual and motor task goals. *Journal of Cognitive Neuroscience* 23(11): 3318–3330.

De Bruijn, M. J., & Bender, M. (2018). Olfactory cues are more effective than visual cues in experimentally triggering autobiographical memories. *Memory* 26(4): 547–558.

De Luca, R., & Botelho, D. (2021). The unconscious perception of smells as a driver of consumer responses: A framework integrating the emotion-cognition approach to scent marketing. *AMS Review* 11(1–2): 145–161.

Decety, J., & Jackson, P. L. (2004). The functional architecture of human empathy. *Behavioral and Cognitive Neuroscience Reviews* 3(2): 71–100.

Denburg, N. L., & Hedgcock, W. M. (2015). Age-associated executive dysfunction, the prefrontal cortex, and complex decision making. In *Aging and Decision Making* (pp. 79–101). Academic Press.

Deng, Y., Chang, L., Yang, M., Huo, M., & Zhou, R. (2016). Gender differences in emotional response: Inconsistency between experience and expressivity. *PloS one* 11(6): e0158666.

Derke, F., Filipović-Grčić, L., Raguž, M., Lasić, S., Orešković, D., & Demarin, V. (2023). *Neuroaesthetics: How We Like What We Like. In Mind, Brain and Education* (pp. 1–12). Cham: Springer International Publishing.

Derntl, B., Kryspin-Exner, I., Fernbach, E., Moser, E., & Habel, U. (2008). Emotion recognition accuracy in healthy young females is associated with cycle phase. *Hormones and Behavior* 53(1): 90–95.

Deterding, S., Sicart, M., Nacke, L., O'Hara, K., & Dixon, D. (2011). Gamification: Using game-design elements in non-gaming contexts. In CHI'11 extended abstracts on human factors in computing systems (pp. 2425–2428).

Devlin, J., Chang, M. W., Lee, K., & Toutanova, K. (2018). Bert: Pre-training of deep bidirectional transformers for language understanding. *arXiv preprint arXiv:1810.04805.*

Dias, B., Ressler, K. (2014). Parental olfactory experience influences behavior and neural structure in subsequent generations. *Nat Neurosci* 17: 89–96. **https://doi.org/10.1038/nn.3594.**

Dings, R., & Newen, A. (2023). Constructing the past: The relevance of the narrative self in modulating episodic memory. *Review of Philosophy and Psychology* 14(1): 87–112.

Domes, G., Schulze, L., Böttger, M., Grossmann, A., Hauenstein, K., Wirtz, P. H., . . . & Herpertz, S. C. (2010). The neural correlates of sex differences in emotional reactivity and emotion regulation. *Human Brain Mapping* 31(5): 758–769.

Dosovitskiy, A., Beyer, L., Kolesnikov, A., Weissenborn, D., Zhai, X., Unterthiner, T., . . . & Houlsby, N. (2020). An image is worth 16x16 words: Transformers for image recognition at scale. *arXiv preprint arXiv:2010.11929.*

Doty, R. L., & Kamath, V. (2014). The influences of age on olfaction: a review. *Frontiers in Psychology* 5: 72845.

Dowd, T. (2015). *Storytelling across Worlds: Transmedia for Creatives and Producers.* CRC Press.

Dunn, T. L., & Risko, E. F. (2019). Understanding the cognitive miser: Cue-utilization in effort-based decision making. *Acta Psychologica* 198: 102863.

Eijlers E, Boksem MAS and Smidts A (2020) Measuring neural arousal for advertisements and its relationship with advertising success. *Frontiers in Neuroscience* 14: 736. doi: 10.3389/fnins.2020.00736

Eisenberger, N. I., Lieberman, M. D., & Williams, K. D. (2003). Does rejection hurt? An FMRI study of social exclusion. *Science* 302(5643): 290–292.

Eisend, M. (2022). The influence of humor in advertising: Explaining the effects of humor in two-sided messsages. *Psychology & Marketing* 39(5): 962–973.

Elliot, A. J., & Maier, M. A. (2014). Color psychology: Effects of perceiving color on psychological functioning in humans. *Annual Review of Psychology* 65: 95–120.

Elyada, Y. M., & Mizrahi, A. (2015). Becoming a mother—circuit plasticity underlying maternal behavior. *Current Opinion in Neurobiology* 35: 49–56.

Erickson, K. I., Gildengers, A. G., & Butters, M. A. (2013). Physical activity and brain plasticity in late adulthood. *Dialogues in Clinical Neuroscience* 15(1): 99–108.

Ericsson, K. A., Krampe, R. T., & Tesch-Römer, C. (1993). The role of deliberate practice in the acquisition of expert performance. *Psychological Review* 100(3): 363–406.

Escalas, J. E., & Bettman, J. R. (2005). Self-construal, reference groups, and brand meaning. *Journal of Consumer Research* 32(3): 378–389.

Fatourechi, M., et al. (2007). A Self-paced and Calibration-less SSVEP-based Brain–Computer Interface Speller. *IEEE Transactions on Neural Systems and Rehabilitation Engineering* 18(2): 127–133.

Fauconnier, G., & Turner, M. (2003). Conceptual blending, form and meaning. *Recherches en Communication* 19: 57–86.

Finset, A. (2014). Talk-in-interaction and neuropsychological processes. *Scandinavian Journal of Psychology* 55(3): 212–218.

Foulkes, L., & Blakemore, S. J. (2016). Is there heightened sensitivity to social reward in adolescence? *Current Opinion in Neurobiology* 40, 81–85.

Frater, J., & Hawley, J. M. (2018). A hand-crafted slow revolution: Co-designing a new genre in the luxury world. *Fashion, Style & Popular Culture* 5(3): 299–311.

Friston, K. (2005). A theory of cortical responses. Philosophical transactions of the Royal Society of London. Series B. *Biological Sciences* 360(1456): 815–836.

Fu, J., Tan, L. K., Li, N. P., & Wang, X. (XiaoTian). (2024). Imprinting-like effects of early adolescent music. *Psychology of Music* 52(1): 38–58. **https://doi.org/10.1177/030573562 31156201**

Furnham, A., & Boo, H. C. (2011). A literature review of the anchoring effect. *The Journal of Socio-Economics* 40(1): 35–42.

Gallegos, I. O., Rossi, R. A., Barrow, J., Tanjim, M. M., Yu, T., Deilamsalehy, H., . . . & Dernoncourt, F. (2024). Self-Debiasing Large Language Models: Zero-Shot Recognition and Reduction of Stereotypes. *arXiv preprint arXiv:2402.01981.*

Gallucci, M., & Perugini, M. (2003). Information seeking and reciprocity: A transformational analysis. *European Journal of Social Psychology, 33*(4): 473–495.

Galtress, T., & Kirkpatrick, K. (2010). The role of the nucleus accumbens core in impulsive choice, timing, and reward processing. *Behavioral neuroscience*, 124(1): 26.

Ganisasmara, N. S., & Mani, L. (2020). The Effect of Celebrity Endorsement, Review, and Viral Marketing on Purchase Decision of X Cosmetics. *Solid State Technology* 63(5): 9679–9697.

Garg, N., Sethupathy, A., Tuwani, R., Nk, R., Dokania, S., Iyer, A., . . . & Bagler, G. (2018). FlavorDB: a database of flavor molecules. *Nucleic Acids Research* 46(D1): D1210–D1216.

Gau, W. B. (2019). A reflection on marketing 4.0 from the perspective of senior citizens' communities of practice. *Sage Open* 9(3): 2158244019867859.

Geary, D. C. (1998). *Male, Female: The Evolution of Human Sex Differences* (p. 110). Washington, DC: American Psychological Association.

Geniole, S. N., & Carré, J. M. (2018). Human social neuroendocrinology: Review of the rapid effects of testosterone. Hormones and behavior, 104, 192–205.

Gerrits, R., Verhelst, H., & Vingerhoets, G. (2020). Mirrored brain organization: Statistical anomaly or reversal of hemispheric functional segregation bias? *Proceedings of the National Academy of Sciences,* 117(25): 14057–14065.

Gerten, J., Zürn, M. K., & Topolinski, S. (2022). The price of predictability: estimating inconsistency premiums in social interactions. *Personality and Social Psychology Bulletin* 48(2): 183–202.

Gigerenzer, G., & Gaissmaier, W. (2011). Heuristic decision making. *Annual Review of Psychology* 62: 451–482.

Gilligan, I. (2019). *Climate, Clothing, and Agriculture in Prehistory: Linking Evidence, Causes, and Effects.* Cambridge University Press.

Glaser, M., & Reisinger, H. (2022). Don't lose your product in story translation: How product–story link in narrative advertisements increases persuasion. *Journal of Advertising* 51(2): 188–205.

Gobbini, M. I., Koralek, A. C., Bryan, R. E., Montgomery, K. J., & Haxby, J. V. (2007). Two takes on the social brain: A comparison of theory of mind tasks. *Journal of Cognitive Neuroscience* 19(11): 1803–1814.

Google, "PaLM 2," **https://ai.google/discover/palm2/**

Google, "Introducing Gemini: our largest and most capable AI model," 06 December 2023. [Online]. Available: **https://blog.google/technology/ai/google-gemini-ai/** (accessed 21 December 2023).

Google, "Exploring Transfer Learning with T5: the Text-To-Text Transfer Transformer," 4 February 2020. [Online]. Available: **https://blog.research.google/2020/02/exploring-transfer-learning-with-t5.html** (accessed 21 December 2023).

Google, "Pathways Language Model (PaLM): Scaling to 540 Billion Parameters for Breakthrough Performance," 4 April 2022. **https://blog.research.google/2022/04/pathways-language-model-palm-scaling-to.html** (accessed 21 December 2023).

Gottlieb, J., & Oudeyer, P. Y. (2018). Towards a neuroscience of active sampling and curiosity. *Nature Reviews Neuroscience*, 19(12): 758–770.

Grahn, J. A., & Brett, M. (2007). Rhythm and beat perception in motor areas of the brain. *Journal of Cognitive Neuroscience*, 19(5): 893–906.

Graybiel, A. M., & Grafton, S. T. (2015). The striatum: where skills and habits meet. Cold Spring Harbor perspectives in biology, 7(8): a021691.

Greeff, O. G. (2020). Serotonin in the elderly. *South African General Practitioner,* 1(2): 79–81.

Green, J. D., Reid, C. A., Kneuer, M. A., & Hedgebeth, M. V. (2023). The proust effect: Scents, food, and nostalgia. *Current Opinion in Psychology* 50: 101562.

Gretchen Rubin & Oracle Fusion Cloud Customer Experience, 2022. Global Report: 45% of People Have Not Felt True Happiness for More Than Two Years.

Gur, R. C., Turetsky, B. I., Matsui, M., Yan, M., Bilker, W., Hughett, P., & Gur, R. E. (1999). Sex differences in brain gray and white matter in healthy young adults: correlations with cognitive performance. *Journal of Neuroscience* 19(10): 4065–4072.

Gvili, Y., Levy, S., & Zwilling, M. (2018). The sweet smell of advertising: The essence of matching scents with other ad cues. *International Journal of Advertising* 37(4): 568–590.

Hagestad, G. O. (2018). Interdependent lives and relationships in changing times: A life-course view of families and aging. *Invitation to the Life Course* (pp. 135–159). Routledge.

Hamid, A. A., Pettibone, J. R., Mabrouk, O. S., Hetrick, V. L., Schmidt, R., Vander Weele, C. M., . . . & Berke, J. D. (2016). Mesolimbic dopamine signals the value of work. *Nature Neuroscience* 19(1): 117–126.

Hampson, E. (1990). Variations in sex-related cognitive abilities across the menstrual cycle. *Brain and Cognition* 14(1): 26–43.

Harmon-Jones, E., Gable, P. A., & Peterson, C. K. (2010). The role of asymmetric frontal cortical activity in emotion-related phenomena: A review and update. *Biological Psychology* 84(3): 451–462.

Hartmann, J., Heitmann, M., Schamp, C., & Netzer, O. (2021). The power of brand selfies. *Journal of Marketing Research* 58(6): 1159–1177.

Heinrich, A., Gagne, J. P., Viljanen, A., Levy, D. A., Ben-David, B. M., & Schneider, B. A. (2016). Effective communication as a fundamental aspect of active aging and well-being: paying attention to the challenges older adults face in noisy environments. *Social Inquiry into Well-Being* 2(1).

Helfrich, R. F., Mander, B. A., Jagust, W. J., Knight, R. T., & Walker, M. P. (2018). Old brains come uncoupled in sleep: slow wave-spindle synchrony, brain atrophy, and forgetting. *Neuron* 97(1): 221–230.

Hensler, J. G. (2010). Serotonin in mood and emotion. In *Handbook of Behavioral Neuroscience* (Vol. 21, pp. 367–378). Elsevier.

Herrmann, C. S., Strüber, D., Helfrich, R. F., & Engel, A. K. (2016). EEG oscillations: from correlation to causality. *International Journal of Psychophysiology* 103: 12–21.

Hikida, T., Morita, M., & Macpherson, T. (2016). Neural mechanisms of the nucleus accumbens circuit in reward and aversive learning. *Neuroscience Research* 108: 1–5.

Hirnstein, M., Stuebs, J., Moè, A., & Hausmann, M. (2023). Sex/gender differences in verbal fluency and verbal-episodic memory: a meta-analysis. *Perspectives on Psychological Science* 18(1): 67–90.

Hoba, S., Fink, G. R., Zeng, H., & Weidner, R. (2022). View normalization of object size in the right parietal cortex. *Vision* 6(3): 41.

Hoekzema, E. et al. (2017). Pregnancy leads to long-lasting changes in human brain structure. *Nature Neuroscience* 20(2): 287–296.

Holroyd, C. B., Nieuwenhuis, S., Mars, R. B., & Coles, M. G. (2004). Anterior cingulate cortex, selection for action, and error processing. *Cognitive Neuroscience of Attention* 219–231.

Horoszko, N., Moskowitz, D., & Moskowitz, H. (2018). *Understanding the Marketing Exceptionality of Prestige Perfumes*. Routledge.

Hsu, L., & Chen, Y. J. (2020). Neuromarketing, subliminal advertising, and hotel selection: An EEG study. *Australasian Marketing Journal* 28(4): 200–208.

huggingface, "Unit 3. Transformer architectures for audio," **https://huggingface.co/ learn/audio-course/chapter3/introduction** (accessed 21 December 2023).

Huntone, P. (2016). The Benefits of Haptic Feedback in Mobile Phone Camera (master's thesis).

Huron, D. (2011). Why is sad music pleasurable? A possible role for prolactin. *Musicae Scientiae* 15(2): 146–158.

Inzlicht, M., Shenhav, A., & Olivola, C. Y. (2018). The effort paradox: Effort is both costly and valued. *Trends in Cognitive Sciences* 22(4): 337–349.

Ishiguro, H. (2006). Interactive humanoids and androids as ideal interfaces for humans. In *International Conference on Intelligent User Interfaces* (pp. 2–9). ACM.

Itti, L. (2007). Visual salience. *Scholarpedia* 2(9): 3327.

Jacques, P. S., Dolcos, F., & Cabeza, R. (2010). Effects of aging on functional connectivity of the amygdala during negative evaluation: A network analysis of fMRI data. *Neurobiology of Aging* 31(2): 315–327.

Jain, G., Shrivastava, S., Nayakankuppam, D., & Gaeth, G. J. (2021). (The lack of) fluency and perceptions of decision making. *Journal of Marketing Communications* 27(6): 670–684.

Jain, P., & Jain, U. (2016). Study of the effectiveness of advertising jingles. *Advances in Economics and Business Management* 3(5): 596–505.

James, A. N. (Ed.). (2015). *Teaching the Male Brain: How Boys Think, Feel, and Learn in School.* Corwin Press.

Jostmann, N. B., Lakens, D., & Schubert, T. W. (2009). Weight as an embodiment of importance. *Psychological Science* 20(9): 1169–1174.

Jung, Carl G. (1959). *The Archetypes and the Collective Unconscious.* R. F. C. Hull, Trans. Princeton, NJ: Princeton University Press.

Juniper, A. (2011). *Wabi Sabi: The Japanese Art of Impermanence.* Tuttle Publishing.

Juslin, P. N., & Laukka, P. (2003). Communication of emotions in vocal expression and music performance: Different channels, same code? *Psychological Bulletin* 129(5): 770.

Kahneman, D., & Tversky, A. (2013). Prospect theory: An analysis of decision under risk. *Handbook of the Fundamentals of Financial Decision Making: Part I* (pp. 99–127).

Kaplan, J. T., & Iacoboni, M. (2006). Getting a grip on other minds: Mirror neurons, intention understanding, and cognitive empathy. *Social Neuroscience* 1(3–4): 175–183.

Kanwisher, N., McDermott, J., & Chun, M. M. (1997). The fusiform face area: A module in human extrastriate cortex specialized for face perception. *Journal of Neuroscience* 17(11): 4302–4311.

Karpasitis, C., Polycarpou, I., & Kaniadakis, A. (2018, February). The role of music in viral video advertisements. In *ECSM 2018 5th European Conference on Social Media* (pp. 93–100). Academic Conferences and Publishing Limited.

Kaufman, Barbara. (2003). Stories that SELL, stories that TELL. *Journal of Business Strategy* 24: 11–15. 10.1108/02756660310508155.

Keane, M. P., & Thorp, S. (2016). Complex decision making: the roles of cognitive limitations, cognitive decline, and aging. In *Handbook of the Economics of Population Aging* (Vol. 1, pp. 661–709). North-Holland.

Kellaris, J. J. (2018). Music and consumers. *Handbook of Consumer Psychology*, 828–847.

Keller, H. (1927). *My Religion.* Doubleday.

Kerzel, D., & Schönhammer, J. (2013). Salient stimuli capture attention and action. *Attention, Perception, & Psychophysics* 75: 1633–1643.

Kilner, J. M., & Blakemore, S. J. (2007). How does the mirror neuron system change during development? *Developmental Science* 10(5): 524–526.

Kim, H., Wang, K., Cutting, L. E., Willcutt, E. G., Petrill, S. A., Leopold, D. R., . . . & Banich, M. T. (2022). The angular gyrus as a hub for modulation of language-related cortex by distinct prefrontal executive control regions. *Journal of Cognitive Neuroscience,* 34(12): 2275—2296.

Kim, J., Strohbach, C. A., & Wedell, D. H. (2019). Effects of manipulating the tempo of popular songs on behavioral and physiological responses. *Psychology of Music* 47(3): 392–406.

Kim, P., Dufford, A. J., & Tribble, R. C. (2018). Cortical thickness variation of the maternal brain in the first 6 months postpartum: associations with parental self-efficacy. *Brain Structure and Function* 223: 3267–3277.

Kim, S.-G., & Ogawa, S. (2012). Biophysical and physiological origins of blood oxygenation level-dependent fMRI signals. *Journal of Cerebral Blood Flow & Metabolism* 32(7): 1188–1206.

Kleim, J.A., & Jones, T.A. (2008). Principles of experience-dependent neural plasticity: Implications for rehabilitation after brain damage. *Journal of Speech, Language, and Hearing Research* 51(1): S225–S239.

Klein, S. B. (2016). Autonoetic consciousness: Reconsidering the role of episodic memory in future-oriented self-projection. *Quarterly Journal of Experimental Psychology* 69(2): 381–401.

Klinger, E. (1990). *Daydreaming.* Los Angeles: Tarcher.

Klinzing, J. G., Niethard, N., & Born, J. (2019). Mechanisms of systems memory consolidation during sleep. *Nature Neuroscience* 22(10): 1598–1610.

Kluen, L. M., Agorastos, A., Wiedemann, K., & Schwabe, L. (2017). Cortisol boosts risky decision-making behavior in men but not in women. *Psychoneuroendocrinology* 84: 181–189.

Knapp, H. P., & Corina, D. P. (2010). A human mirror neuron system for language: Perspectives from signed languages of the deaf. *Brain and Language* 112(1): 36–43.

Knutson, B., et al. (2007). Neural Predictors of Purchases. *Neuron* 53(1): 147–156.

Kohlhoff, J., Eapen, V., Dadds, M., Khan, F., Silove, D., & Barnett, B. (2017). Oxytocin in the postnatal period: Associations with attachment and maternal caregiving. *Comprehensive Psychiatry* 76: 56–68.

Kosfeld, M., Heinrichs, M., Zak, P. J., Fischbacher, U., & Fehr, E. (2005). Oxytocin increases trust in humans. *Nature* 435(7042): 673–676.

Kosyakovsky, J. (2021). The neural economics of brain aging. *Scientific Reports* 11(1): 12167.

Kotek, H., Dockum, R., & Sun, D. (2023, November). Gender bias and stereotypes in large language models. In *Proceedings of the ACM Collective Intelligence Conference* (pp. 12–24).

Kotz, S. A., Ravignani, A., & Fitch, W. T. (2018). The evolution of rhythm processing. *Trends in Cognitive Sciences* 22(10): 896–910.

Krach, S., Cohrs, J. C., de Echeverría Loebell, N. C., Kircher, T., Sommer, J., Jansen, A., & Paulus, F. M. (2011). Your flaws are my pain: Linking empathy to vicarious embarrassment. *PloS one* 6(4): e18675.

Kramer, Thomas, and Lauren Block. "Nonconscious effects of peculiar beliefs on consumer psychology and choice." *Journal of Consumer Psychology* 21.1 (2011): 101–111.

Krebs, R. M., Heipertz, D., Schuetze, H., & Duzel, E. (2011). Novelty increases the mesolimbic functional connectivity of the substantia nigra/ventral tegmental area (SN/VTA) during reward anticipation: Evidence from high-resolution fMRI. *Neuroimage* 58(2): 647–655.

Krishna, A. (2012). An integrative review of sensory marketing: Engaging the senses to affect perception, judgment and behavior. *Journal of Consumer Psychology* 22(3): 332–351.

Kühn, S., Brick, T. R., Müller, B. C., & Gallinat, J. (2014). Is this car looking at you? How anthropomorphism predicts fusiform face area activation when seeing cars. *PloS one* 9(12): e113885.

Kuno, S., & Kaburaki, E. (1977). Empathy and syntax. *Linguistic Inquiry*, 627–672.

Lacoste-Badie, S., & Droulers, O. (2014). Advertising memory: The power of mirror neurons. *Journal of Neuroscience, Psychology, and Economics* 7(4): 195.

Lakoff, G., & Johnson, M. (1980). *Metaphors We Live By*. University of Chicago Press.

Laksmidewi, D., Susianto, H., & Afiff, A. Z. (2017). Anthropomorphism in advertising: the effect of anthropomorphic product demonstration on consumer purchase intention. *Asian Academy of Management Journal* 22(1).

Lamm, C., & Majdandžić, J. (2015). The role of shared neural activations, mirror neurons, and morality in empathy – A critical comment. *Neuroscience Research* 90: 15–24.

Larson, C. L., Aronoff, J., Sarinopoulos, I. C., & Zhu, D. C. (2009). Recognizing threat: a simple geometric shape activates neural circuitry for threat detection. *Journal of Cognitive Neuroscience* 21(8): 1523–1535.

LeDoux, J. E. (1996). *The Emotional Brain: The Mysterious Underpinnings of Emotional Life*. Simon & Schuster.

LeDoux, J., & Daw, N. D. (2018). Surviving threats: neural circuit and computational implications of a new taxonomy of defensive behaviour. *Nature Reviews Neuroscience* 19(5): 269–282.

Lee, B. P., & Spence, C. (2022). Crossmodal correspondences between basic tastes and visual design features: A narrative historical review. *I-Perception* 13(5). **https://doi.org/10.1177/20416695221127325**

Lembke, A. (2021). *Dopamine Nation: Finding Balance in the Age of Indulgence*. Penguin.

Lent, J. (2021). *The Web of Meaning: Integrating Science and Traditional Wisdom to Find Our Place in the Universe*. New Society Publishers.

Levitin, D. (2020). *The Changing Mind: A Neuroscientist's Guide to Ageing Well*. Penguin UK.

Levy, B. (2009). Stereotype embodiment: A psychosocial approach to aging. *Current Directions in Psychological Science* 2009 Dec 1; 18(6): 332–336. doi: 10.1111/j.1467–8721.2009.01662.x.

Levy, N., Harmon-Jones, C., & Harmon-Jones, E. (2018). Dissonance and discomfort: Does a simple cognitive inconsistency evoke a negative affective state? *Motivation Science* 4(2): 95.

Lewis, M. (2015). *The Biology of Desire: Why Addiction Is Not a Disease*. Public Affairs Books.

Liang, H., Bressler, S. L., Ding, M., Truccolo, W. A., & Nakamura, R. (2002). Synchronized activity in prefrontal cortex during anticipation of visuomotor processing. *Neuroreport* 13(16): 2011–2015.

Liao, H. I., Yeh, S. L., & Shimojo, S. (2011). Novelty vs. familiarity principles in preference decisions: task-context of past experience matters. *Frontiers in Psychology* 2: 7703.

Lieberman, D. E., Kistner, T. M., Richard, D., Lee, I. M., & Baggish, A. L. (2021). The active grandparent hypothesis: Physical activity and the evolution of extended human health-spans and lifespans. *Proceedings of the National Academy of Sciences* 118(50): e2107621118.

Lipton, D. M., Gonzales, B. J., & Citri, A. (2019). Dorsal striatal circuits for habits, compulsions and addictions. *Frontiers in Systems Neuroscience* 13: 28.

List, C., & Kipp, M. (2019, August). Is bigger better? A Fitts' law study on the impact of display size on touch performance. In *IFIP Conference on Human-Computer Interaction* (pp. 669–678). Cham: Springer International Publishing.

Long, N. M., & Kahana, M. J. (2015). Successful memory formation is driven by contextual encoding in the core memory network. *NeuroImage* 119, 332–337.

Lu, J., Xue, G., & Dong, Q. (2016). Emoji: Communicating emotions in the digital era. *Global Media and Communication* 12(3): 299–302.

Lui, K. F., Yip, K. H., & Wong, A. C. (2021). Gender differences in multitasking experience and performance. *Quarterly Journal of Experimental Psychology* 74(2): 344–362.

Lunardo, R., & Livat, F. (2016). Congruency between colour and shape of the front labels of wine: effects on fluency and aroma and quality perceptions. *International Journal of Entrepreneurship and Small Business* 29(4): 528–541.

Maaike J. de Bruijn & Michael Bender (2018) Olfactory cues are more effective than visual cues in experimentally triggering autobiographical memories. *Memory* 26(4): 547–558, DOI: 10.1080/09658211.2017.1381744

Maccoby, E. E. (1998). *The Two Sexes: Growing Up Apart, Coming Together* (Vol. 4). Harvard University Press.

Maimaran, M., & Wheeler, S. C. (2008). Circles, squares, and choice: The effect of shape arrays on uniqueness and variety seeking. *Journal of Marketing Research* 45(6): 731–740.

MAKEUSEOF, "GPT-1 to GPT-4: Each of OpenAI's GPT Models Explained and Compared," April 2023. **https://www.makeuseof.com/gpt-models-explained-and-compared/** (accessed 21 December 2023).

Mäntylä, T. (2013). Gender differences in multitasking reflect spatial ability. *Psychological Science, 24*(4): 514–520.

Mar, R. A. (2011). The neural bases of social cognition and story comprehension. *Annual Review of Psychology* 62: 103–134.

Marceau, K., Zahn-Waxler, C., Shirtcliff, E. A., Schreiber, J. E., Hastings, P., & Klimes-Dougan, B. (2015). Adolescents', mothers', and fathers' gendered coping strategies during conflict: Youth and parent influences on conflict resolution and psychopathology. *Development and Psychopathology* 27(4pt1): 1025–1044.

Mason, A., Farrell, S., Howard-Jones, P., & Ludwig, C. J. (2017). The role of reward and reward uncertainty in episodic memory. *Journal of Memory and Language* 96: 62–77.

Mathur, P., Chun, H. H., & Maheswaran, D. (2016). Consumer mindsets and self-enhancement: Signaling versus learning. *Journal of Consumer Psychology* 26(1): 142–152.

Mayer, R. E. & Moreno, R. (2003). Nine ways to reduce cognitive load in multimedia learning. *Educational Psychologist* 38(1): 43–52.

McAlexander, J. H., Schouten, J. W., & Koenig, H. F. (2002). Building brand community. *Journal of Marketing* 66(1): 38–54.

McCaffrey, T. (2012). Innovation relies on the obscure: A key to overcoming the classic problem of functional fixedness. *Psychological Science* 23(3): 215–218.

McClean, E. J., Martin, S. R., Emich, K. J., & Woodruff, C. T. (2018). The social consequences of voice: An examination of voice type and gender on status and subsequent leader emergence. *Academy of Management Journal* 61(5): 1869–1891.

McClure, E. B. (2000). A meta-analytic review of sex differences in facial expression processing and their development in infants, children, and adolescents. *Psychological Bulletin* 126(3): 424.

McClure SM, Li J, Tomlin D, Cypert KS, Montague LM, Montague PR. (2004). Neural correlates of behavioral preference for culturally familiar drinks. *Neuron* Oct 14; 44(2): 379–387. Doi: 10.1016/j.neuron.2004.09.019.

McGaugh, J. L. (2003). *Memory and Emotion: The Making of Lasting Memories.* Columbia University Press.

McNeil, P., & Riello, G. (2016). *Luxury: A Rich History.* Oxford University Press.

Medina, J., & Workman, J. L. (2020). Maternal experience and adult neurogenesis in mammals: implications for maternal care, cognition, and mental health. *Journal of Neuroscience Research* 98(7): 1293–1308.

Mefoh, P. C., Nwoke, M. B., Chukwuorji, J. C., & Chijioke, A. O. (2017). Effect of cognitive style and gender on adolescents' problem solving ability. *Thinking Skills and Creativity* 25, 47–52.

Menon, R., Süß, T., de Moura Oliveira, V. E., Neumann, I. D., & Bludau, A. (2022). Neurobiology of the lateral septum: regulation of social behavior. *Trends in Neurosciences* 45(1): 27–40.

META, "Introducing Llama 2," **https://ai.meta.com/llama/**

Miller, E. K., & Cohen, J. D. (2001). An integrative theory of prefrontal cortex function. *Annual Review of Neuroscience* 24: 167–202. **https://doi.org/10.1146/annurev.neuro.24.1.167**

Miller, G.A. (1956). The magical number seven, plus or minus two: some limits on our capacity for processing information. *Psychological Review* 63(2): 81–97.

Minsky, L., & Fahey, C. (2017). *Audio Branding: Using Sound to Build Your Brand.* Kogan Page Publishers.

Montgomery, A. L., & Smith, M. D. (2009). Prospects for personalized marketing as information technology improves. *Journal of Interactive Marketing* 23(2): 130–137.

Moody, E. J., McIntosh, D. N., Mann, L. J., & Weisser, K. R. (2007). More than mere mimicry? The influence of emotion on rapid facial reactions to faces. *Emotion* 7(2): 447.

Moore, A. K., & Miller, R. J. (2020). Video storytelling in the classroom: The role of narrative transportation. *Journal of Nursing Education* 59(8): 470–474.

Moore, B. C. (2014). Psychoacoustics. *Handbook of Acoustics.* 475–517. Springer.

Moran, J. M., Wig, G. S., Adams, R. B., Janata, P., & Kelley, W. M. (2004). Neural correlates of humor detection and appreciation. *NeuroImage* 21(3): 1055–1060.

Moreno, S., & Bidelman, G. M. (2014). Examining neural plasticity and cognitive benefit through the unique lens of musical training. *Hearing Research* 308, 84–97.

Moser, M. B., Rowland, D. C., & Moser, E. I. (2015). Place cells, grid cells, and memory. *Cold Spring Harbor Perspectives in Biology* 7(2): a021808.

Muniz Jr, A. M., & O'Guinn, T. C. (2001). Brand community. *Journal of Consumer Research* 27(4): 412–432.

Munro, C. A., McCaul, M. E., Wong, D. F., Oswald, L. M., Zhou, Y., Brasic, J., . . . & Wand, G. S. (2006). Sex differences in striatal dopamine release in healthy adults. *Biological Psychiatry* 59(10): 966–974.

Murillo-Cuesta, S., Rodríguez-de La Rosa, L., Cediel, R., Lassaletta, L., & Varela-Nieto, I. (2011). The role of insulin-like growth factor-I in the physiopathology of hearing. *Frontiers in Molecular Neuroscience* 4: 11.

Murphy, S., Melandri, E., & Bucci, W. (2021). The effects of story-telling on emotional experience: An experimental paradigm. *Journal of Psycholinguistic Research* 50: 117–142.

Musil, C. M., Zauszniewski, J. A., Givens, S. E., Henrich, C., Wallace, M., Jeanblanc, A., & Burant, C. J. (2019). Resilience, resourcefulness, and grandparenting. In B. Hayslip, Jr. & C. A. Fruhauf (Eds.), *Grandparenting: Influences on the Dynamics of Family Relationships* (pp. 233–250). Springer Publishing. **https://doi .org/10.1891/9780826149855.0014**

Nadányiová, M. (2018). The brand building through viral marketing on social networks and its perception by different consumers' generations. *Marketing Identity* 6(1/1): 441–450.

Naranjo-Hernández, D., Reina-Tosina, J., & Roa, L. M. (2020). Sensor technologies to manage the physiological traits of chronic pain: A review. *Sensors* 20(2): 365.

Nehls, S., Losse, E., Enzensberger, C., Frodl, T., & Chechko, N. (2024). Time-sensitive changes in the maternal brain and their influence on mother-child attachment. *Translational Psychiatry* 14(1): 1–9.

Nguyen, D., Nguyen, P., Le, H., Do, K., Venkatesh, S., & Tran, T. (2023, June). Memory-augmented theory of mind network. *Proceedings of the AAAI Conference on Artificial Intelligence* 37(10): 11630–11637.

Nikolinakou, A., & King, K. W. (2018). Viral video ads: Examining motivation triggers to sharing. *Journal of Current Issues & Research in Advertising* 39(2): 120–139.

O'Connell, M. (2018). *And Now We Have Everything: On Motherhood Before I Was Ready*. Little, Brown.

o'g'li, A. A. A., & Zamonbekovich, Z. A. (2023). Visual marketing. *Best Journal of Innovation in Science, Research and Development* 2(5): 371–374.

O'Mara, S. (2017). *A Brain for Business–A Brain for Life: How Insights from Behavioural and Brain Science Can Change Business and Business Practice for the Better*. Springer.

OpenAI, "GPT-4," **https://openai.com/research/gpt-4** (accessed 21 December 2023).

OpenAI, "GPT-2: 1.5B release," **https://openai.com/research/gpt-2-1-5b-release** (accessed 21 December 2023).

OpenAI, "Introducing ChatGPT," **https://openai.com/blog/chatgpt** (accessed 21 December 2023).

OpenAI, "Tokenizer," **https://platform.openai.com/tokenizer** (accessed 21 December 2023).

OpenAI, "Aligning language models to follow instructions," **https://openai.com/ research/instruction-following** (accessed 21 December 2023).

OpenAI, "Fine-tuning GPT-2 from human preferences," **https://openai.com/research/ fine-tuning-gpt-2** (accessed 21 December 2023).

OpenAI, "Improving mathematical reasoning with process supervision," **https://openai .com/research/improving-mathematical-reasoning-with-process-supervision** (accessed 21 December 2023).

OpenAI, "DALL·E: Creating images from text," 5 January 2021. **https://openai.com/ research/dall-e** (accessed 21 December 2023).

Osei-Frimpong, K., Donkor, G., & Owusu-Frimpong, N. (2019). The impact of celebrity endorsement on consumer purchase intention: An emerging market perspective. *Journal of Marketing Theory and Practice* 27(1): 103–121.

Palmer, S., & Rock, I. (1994). Rethinking perceptual organization: The role of uniform connectedness. *Psychonomic Bulletin & Review* 1(1): 29–55.

Palumbo, L., Ruta, N., & Bertamini, M. (2015). Comparing angular and curved shapes in terms of implicit associations and approach/avoidance responses. *PloS one* 10(10): e0140043.

Parker, J. (2022). Delivering return on imagination: a framework for creativity in developing advertising message strategy. Doctoral dissertation, Macquarie University.

Patton, A. P., & Hastings, M. H. (2018). The suprachiasmatic nucleus. *Current Biology* 28(15): R816–R822.

Pavlov, I. P. (1927). *Conditioned Reflexes*. Oxford: Oxford University Press.

Pelham, B. W., Carvallo, M., & Jones, J. T. (2005). Implicit egotism. *Current Directions in Psychological Science* 14(2): 106–110.

Peterson, E., Wallenberg, R., & Källström, J. (2017). Gendered Storytelling: A normative evaluation of gender differences in terms of decoding a message or theme in storytelling. Dissertation from Jönköping University, Jönköping International Business School.

Phelps, E. A. (2004). Human emotion and memory: Interactions of the amygdala and hippocampal complex. *Current Opinion in Neurobiology* 14(2): 198–202.

Phillips, C. (2017). Lifestyle modulators of neuroplasticity: how physical activity, mental engagement, and diet promote cognitive health during aging. *Neural plasticity,* 2017.

Pool, E., Brosch, T., Delplanque, S., & Sander, D. (2016). Attentional bias for positive emotional stimuli: A meta-analytic investigation. *Psychological Bulletin* 142(1): 79.

Potter, M. C., Wyble, B., Hagmann, C. E., & McCourt, E. S. (2014). Detecting meaning in RSVP at 13 ms per picture. *Attention, Perception, & Psychophysics* 76(2): 270–279.

Pozharliev, R., Verbeke, W. J., Van Strien, J. W., & Bagozzi, R. P. (2015). Merely being with you increases my attention to luxury products: Using EEG to understand consumers' emotional experience with luxury branded products. *Journal of Marketing Research* 52(4): 546–558.

Price, A. (2017). *He's Not Lazy: Empowering Your Son to Believe in Himself*. Union Square & Co.

Putkinen, V., Nazari-Farsani, S., Seppälä, K., Karjalainen, T., Sun, L., Karlsson, H. K., ... & Nummenmaa, L. (2021). Decoding music-evoked emotions in the auditory and motor cortex. *Cerebral Cortex* 31(5): 2549–2560.

Putrevu, S. (2001). Exploring the origins and information processing differences between men and women: Implications for advertisers. *Academy of Marketing Science Review* 10(1): 1–14.

Radesky, J. S., Kistin, C., Eisenberg, S., Gross, J., Block, G., Zuckerman, B., & Silverstein, M. (2016). Parent perspectives on their mobile technology use: The excitement and exhaustion of parenting while connected. *Journal of Developmental & Behavioral Pediatrics* 37(9): 694–701.

Radford, A., Narasimhan, K., Salimans, T., & Sutskever, I. (2018). Improving language understanding by generative pre-training. **https://s3-us-west-2.amazonaws.com/openai-assets/research-covers/language-unsupervised/language_understanding_paper.pdf**.

Rafal Ohme, Michal Matukin & Beata Pacula-Lesniak (2011. Biometric measures for interactive advertising research. *Journal of Interactive Advertising* 11(2): 60–72, DOI: 10.1080/15252019.2011.10722185

Raffel, C., Shazeer, N., Roberts, A., Lee, K., Narang, S., Matena, M., ... & Liu, P. J. (2020). Exploring the limits of transfer learning with a unified text-to-text transformer. *Journal of machine learning research* 21(140): 1–67.

Railton, P. (2017). Moral learning: Conceptual foundations and normative relevance. *Cognition* 167: 172–190.

Ramachandran, V. S., & Hirstein, W. (1999). The science of art: A neurological theory of aesthetic experience. *Journal of Consciousness Studies* 6(6–7): 15–51.

Ramesh, A., Dhariwal, P., Nichol, A., Chu, C., & Chen, M. (2022). Hierarchical text-conditional image generation with clip latents. *arXiv preprint arXiv:2204.06125*, 1(2): 3.

Rapaille, C. (2015). *The Global Code: How a New Culture of Universal Values Is Reshaping Business and Marketing*. Macmillan.

Rauschnabel, P. A. (2021). Augmented reality is eating the real-world! The substitution of physical products by holograms. *International Journal of Information Management* 57: 102279.

Rich, E. L., Stoll, F. M., & Rudebeck, P. H. (2018). Linking dynamic patterns of neural activity in orbitofrontal cortex with decision making. *Current Opinion in Neurobiology* 49: 24–32.

Rifqiya, A., & Nasution, R. A. (2016). Sensory marketing: The effect of tactile cue on product packaging towards perceived novelty and perceived likeability. *Journal of Business and Management* 5(3).

Rimkute, J., Moraes, C., & Ferreira, C. (2016). The effects of scent on consumer behaviour. *International Journal of Consumer Studies* 40(1): 24–34.

Rizzolatti, G., & Craighero, L. (2004). The mirror-neuron system. Annual Review of *Neuroscience* 27: 169–192. **https://doi.org/10.1146/annurev.neuro.27.070203.144230**

Rizzolatti, G., & Sinigaglia, C. (2016). The mirror mechanism: a basic principle of brain function. *Nature Reviews Neuroscience* 17(12): 757–765.

Rolls, E. T. (2019). Taste and smell processing in the brain. *Handbook of Clinical Neurology* 164: 97–118.

Rommerud, A. (2016). Music therapy for aphasia – how can music help people with aphasia reclaim speech? Master's thesis.

Rosenholtz, R., Huang, J., & Ehinger, K. A. (2012). Rethinking the role of top-down attention in vision: Effects attributable to a lossy representation in peripheral vision. *Frontiers in Psychology* 3: 17385.

Rutherford, H. J., Byrne, S. P., Crowley, M. J., Bornstein, J., Bridgett, D. J., & Mayes, L. C. (2018). Executive functioning predicts reflective functioning in mothers. *Journal of Child and Family Studies* 27: 944–952.

Sanders, J., & Van Krieken, K. (2018). Exploring narrative structure and hero enactment in brand stories. *Frontiers in Psychology* 9: 411248.

Savage, B. M., Lujan, H. L., Thipparthi, R. R., & DiCarlo, S. E. (2017). Humor, laughter, learning, and health! A brief review. *Advances in Physiology Education* 41(3): 341–347.

Schnall, S., Roper, J., & Fessler, D. M. (2010). Elevation leads to altruistic behavior. *Psychological Science* 21(3): 315–320.

Schubert, T. W. (2005). Your highness: vertical positions as perceptual symbols of power. *Journal of Personality and Social Psychology* 89(1): 1.

Schultz, W. (2007). Behavioral dopamine signals. *Trends in Neurosciences* 30(5): 203–210. **https://doi.org/10.1016/j.tins.2007.03.007**

Schultz, W. (2007). Multiple dopamine functions at different time courses. *Annual Review of Neuroscience* 30: 259–288.

Schwartz, S. J., Kim, S. Y., Whitbourne, S. K., Zamboanga, B. L., Weisskirch, R. S., Forthun, L. F., . . . & Luyckx, K. (2013). Converging identities: dimensions of acculturation and personal identity status among immigrant college students. *Cultural Diversity and Ethnic Minority Psychology* 19(2): 155.

Schwarzer, R., & Luszczynska, A. (2008). Self efficacy. *Handbook of Positive Psychology Assessment* 2(0): 7–217.

Selemon, L. D., & Zecevic, N. (2015). Schizophrenia: a tale of two critical periods for prefrontal cortical development. *Translational Psychiatry* 5(8): e623–e623.

Selva, J., Johansen, A. S., Escalera, S., Nasrollahi, K., Moeslund, T. B., & Clapés, A. (2023). Video transformers: A survey. *IEEE Transactions on Pattern Analysis and Machine Intelligence*.

Seubert, J., Rea, A. F., Loughead, J., & Habel, U. (2009). Mood induction with olfactory stimuli reveals differential affective responses in males and females. *Chemical Senses,* 34(1): 77–84.

Shamay-Tsoory, S. G., & Abu-Akel, A. (2016). The social salience hypothesis of oxytocin. *Biological Psychiatry* 79(3): 194–202.

Sharma, Sukanya, et al. (2020) Social media activities and its influence on customer-brand relationship: an empirical study of apparel retailers' activity in India. *Journal of Theoretical and Applied Electronic Commerce Research* 16(4): 602–617.

Sheldon, S., & Levine, B. (2016). The role of the hippocampus in memory and mental construction. *Annals of the New York Academy of Sciences* 1369(1): 76–92.

Shin, M., Back, K. J., Lee, C. K., & Lee, Y. S. (2022). The loyalty program for our self-esteem: The role of collective self-esteem in luxury hotel membership programs. *Cornell Hospitality Quarterly* 63(1): 19–32

Siegel, D. J. (2015). *Brainstorm: The Power and Purpose of the Teenage Brain*. Penguin.

Silver, D., Huang, A., Maddison, C. J., et al. (2016). Mastering the game of Go with deep neural networks and tree search. *Nature* 529(7587): 484–489.

Simmonds, G., Woods, A. T., & Spence, C. (2018). 'Show me the goods': Assessing the effectiveness of transparent packaging vs. product imagery on product evaluation. *Food Quality and Preference* 63: 18–27.

Simon, D., Becker, M. P., Mothes-Lasch, M., Miltner, W. H., & Straube, T. (2014). Effects of social context on feedback-related activity in the human ventral striatum. *NeuroImage* 99: 1–6.

Singer, T., & Lamm, C. (2009). The social neuroscience of empathy. *Annals of the New York Academy of Sciences* 1156(1): 81–96.

Sirgy, M. J., & Sirgy, M. J. (2020). Positive balance at the emotional level: Hedonic well-being. *Positive Balance: A Theory of Well-Being and Positive Mental Health*: 41–52.

Sisto, D. (2021). *Remember Me: Memory and Forgetting in the Digital Age*. John Wiley & Sons.

Sjostrom, T., Corsi, A. M., & Lockshin, L. (2016). What characterises luxury products? A study across three product categories. *International Journal of Wine Business Research* 28(1): 76–95.

Slater, M., & Sanchez-Vives, M. V. (2016). Enhancing our lives with immersive virtual reality. *Frontiers in Robotics and AI* 3: 74.

Snow, S., & Lazauskas, J. (2018). *The Storytelling Edge: How to Transform Your Business, Stop Screaming into the Void, and Make People Love You*. John Wiley & Sons.

Soch, J., Deserno, L., Assmann, A., Barman, A., Walter, H., Richardson-Klavehn, A., & Schott, B. H. (2017). Inhibition of information flow to the default mode network during self-reference versus reference to others. *Cerebral Cortex* 27(8): 3930–3942.

Soltani, M., & Knight, R. T. (2000). Neural origins of the P300. *Critical Reviews™ in Neurobiology* 14(3–4).

Som, A., & Blanckaert, C. (2015). *The Road to Luxury: The Evolution, Markets, and Strategies of Luxury Brand Management*. John Wiley & Sons.

Spence, C. (2016). Multisensory packaging design: Color, shape, texture, sound, and smell. In *Integrating the Packaging and Product Experience in Food and Beverages* (ed. P. Burgess) (1–22). Woodhead Publishing Series in Food Science, Technology and Nutrition.

Spence, C., Reinoso-Carvalho, F., Velasco, C., & Wang, Q. J. (2019). Extrinsic auditory contributions to food perception & consumer behaviour: An interdisciplinary review. *Multisensory Research* 32(4–5): 275–318.

Spierer, D. K., Petersen, R. A., Duffy, K., Corcoran, B. M., & Rawls-Martin, T. (2010). Gender influence on response time to sensory stimuli. *The Journal of Strength & Conditioning Research* 24(4): 957–963.

Squire, L. R., Genzel, L., Wixted, J. T., & Morris, R. G. (2015). Memory consolidation. *Cold Spring Harbor Perspectives in Biology* 7(8): a021766.

Srivastava, R. K., Fahey, L., & Christensen, H. K. (2001). The resource-based view and marketing: The role of market-based assets in gaining competitive advantage. *Journal of Management* 27(6): 777–802.

Steward, M. (2017). Empathy and the Role of Mirror Neurons. Regis University. **https://core.ac.uk/download/pdf/217368217.pdf**.

Stocchi, L., Wright, M., & Driesener, C. (2016). Why familiar brands are sometimes harder to remember. *European Journal of Marketing* 50(3/4): 621–638.

Streicher, M. C., & Estes, Z. (2015). Touch and go: Merely grasping a product facilitates brand perception and choice. *Applied Cognitive Psychology* 29(3): 350–359.

Sutherland, M. (2020). Advertising and the mind of the consumer: what works, what doesn't and why. Routledge.

Sweller, J. (2011). Cognitive load theory. *Psychology of Learning and Motivation* (Vol. 55, pp. 37–76). Academic Press.

Szameitat, A. J., Hamaida, Y., Tulley, R. S., Saylik, R., & Otermans, P. C. (2015). "Women are better than men"–Public beliefs on gender differences and other aspects in multitasking. *Plos one* 10(10): e0140371.

Tamir, D. I., Bricker, A. B., Dodell-Feder, D., & Mitchell, J. P. (2016). Reading fiction and reading minds: The role of simulation in the default network. *Social Cognitive and Affective Neuroscience* 11(2): 215–224.

Tanhan, F., Özok, H. İ., & Tayiz, V. (2022). Fear of missing out (FoMO): A current review. *Psikiyatride Guncel Yaklasimlar* 14(1): 74–85.

Teichert, M., & Bolz, J. (2018). How senses work together: cross-modal interactions between primary sensory cortices. *Neural Plasticity*, 2018.

Tessitore, A., Hariri, A. R., Fera, F., Smith, W. G., Das, S., Weinberger, D. R., & Mattay, V. S. (2005). Functional changes in the activity of brain regions underlying emotion processing in the elderly. *Psychiatry Research: Neuroimaging* 139(1): 9–18.

Tolkien, J. R. R. (1986). *The Lord of the Ring, Part 1: The Fellowship of the Ring*. Del Rey. Reissue.

Tulving, E. (1972). Episodic and Semantic Memory. In E. Tulving & W. Donaldson (Eds.): *Organization of Memory* (pp. 381–402). New York: Academic Press.

The decoder, "GPT-4 architecture, datasets, costs and more leaked," 11 Jul 2023, **https://the-decoder.com/gpt-4-architecture-datasets-costs-and-more-leaked/** (accessed 21 December 2023).

Twenge, J. M. (2017). *iGen: Why Today's Super-Connected Kids Are Growing Up Less Rebellious, More Tolerant, Less Happy – And Completely Unprepared for Adulthood – And What That Means for the Rest of Us*. Simon and Schuster.

Tyrer, A. E., Levitan, R. D., Houle, S., Wilson, A. A., Nobrega, J. N., & Meyer, J. H. (2016). Increased seasonal variation in serotonin transporter binding in seasonal affective disorder. *Neuropsychopharmacology* 41(10): 2447–2454.

Ushiro, Y., Ogiso, T., Komuro, R., Nahatame, S., Mizugaki, R., Tando, K., & Mikami, Y. (2022). Effects of reading instruction on maintaining coherence of protagonist, temporality, and spatiality in narrative reading. *ARELE: Annual Review of English Language Education in Japan* 33, 65–80.

Uzbay, T. (2020). Importance of Brain Reward System in Neuromarketing. In D. Atli (Ed.), *Analyzing the Strategic Role of Neuromarketing and Consumer Neuroscience* (pp. 1–24). IGI Global. **https://doi.org/10.4018/978–1–7998–3126–6.ch001**

Valkenburg, P. M., & Piotrowski, J. T. (2017). *Plugged In: How Media Attract and Affect Youth*. Yale University Press.

van den Burg, E. H., & Hegoburu, C. (2020). Modulation of expression of fear by oxytocin signaling in the central amygdala: From reduction of fear to regulation of defensive behavior style. *Neuropharmacology* 173, 108130.

Van Kemenade, B. M., Seymour, K., Wacker, E., Spitzer, B., Blankenburg, F., & Sterzer, P. (2014). Tactile and visual motion direction processing in hMT+/V5. *Neuroimage* 84: 420–427.

Van Lieshout, L. L., Vandenbroucke, A. R., Müller, N. C., Cools, R., & De Lange, F. P. (2018). Induction and relief of curiosity elicit parietal and frontal activity. *Journal of Neuroscience* 38(10): 2579–2588.

van Mulukom, V. (2017). Remembering religious rituals: Autobiographical memories of high-arousal religious rituals considered from a narrative processing perspective. *Religion, Brain & Behavior* 7(3): 191–205.

VanRullen, R., Reddy, L., & Koch, C. (2005). Attention-driven discrete sampling of motion perception. *Proceedings of the National Academy of Sciences of the United States of America* 102(14): 5291–5296.

Vaswani, A., Shazeer, N., Parmar, N., Uszkoreit, J., Jones, L., Gomez, A. N., . . . & Polosukhin, I. (2017). Attention is all you need. *Advances in Neural Information Processing Systems* 30.

Verma, P., & Chafe, C. (2021, September). A generative model for raw audio using transformer architectures. In *2021 24th International Conference on Digital Audio Effects (DAFx)* (pp. 230–237). IEEE.

von Gal, A., Boccia, M., Nori, R., Verde, P., Giannini, A. M., & Piccardi, L. (2023). Neural networks underlying visual illusions: An activation likelihood estimation meta-analysis. *NeuroImage*: 120335.

von Restorff, H. (1933). Über die Wirkung von Bereichsbildungen im Spurenfeld. (The effects of field formation in the trace field.) *Psychologische Forschung* 18, 299–342.

Voruganti, L. N., & Awad, A. G. (2007). Role of dopamine in pleasure, reward and subjective responses to drugs. In *Quality of Life Impairment in Schizophrenia, Mood and Anxiety Disorders: New Perspectives on Research and Treatment* (21–31). Springer.

Voutilainen, L., Henttonen, P., Kahri, M., Kivioja, M., Ravaja, N., Sams, M., & Peräkylä, A. (2014). Affective stance, ambivalence, and psychophysiological responses during conversational storytelling. *Journal of Pragmatics* 68: 1–24.

Wang, Z., & Duff, B. R. (2016). All loads are not equal: Distinct influences of perceptual load and cognitive load on peripheral ad processing. *Media Psychology* 19(4): 589–613.

Ward, E. V., Maylor, E. A., Poirier, M., Korko, M., & Ruud, J. C. (2017). A benefit of context reinstatement to recognition memory in aging: the role of familiarity processes. *Aging, Neuropsychology, and Cognition* 24(6): 735–754.

Warren, C., Barsky, A., & McGraw, A. P. (2018). Humor, comedy, and consumer behavior. *Journal of Consumer Research* 45(3): 529–552.

Watts, S. W., Morrison, S. F., Davis, R. P., & Barman, S. M. (2012). Serotonin and blood pressure regulation. *Pharmacological Reviews* 64(2): 359–388.

Weiner, K. S., & Zilles, K. (2016). The anatomical and functional specialization of the fusiform gyrus. *Neuropsychologia* 83: 48–62.

Weis, T., Brechmann, A., Puschmann, S., & Thiel, C. M. (2013). Feedback that confirms reward expectation triggers auditory cortex activity. *Journal of Neurophysiology* 110(8): 1860–1868.

Weizenbaum, J. (1966). ELIZA—a computer program for the study of natural language communication between man and machine. *Communications of the ACM* 9(1): 36–45.

Weston, P. S., Hunter, M. D., Sokhi, D. S., Wilkinson, I. D., & Woodruff, P. W. (2015). Discrimination of voice gender in the human auditory cortex. *NeuroImage* 105: 208–214.

Wheeler, A. (2017). *Designing Brand Identity: An Essential Guide for the Whole Branding Team*. John Wiley & Sons.

White, B. J., & Munoz, D. P. (2011). The superior colliculus. *Oxford Handbook of Eye Movements* 1: 195–213.

Wilson, M. (2002). Six views of embodied cognition. *Psychonomic Bulletin & Review* 9(4): 625–636.

Wingenfeld, K., Wolf, O. T. (2014). Stress, memory, and the hippocampus. *Frontiers of Neurology and Neuroscience* 34: 109–20. doi: 10.1159/000356423.

Wirth, M., Isaacowitz, D. M., & Kunzmann, U. (2017). Visual attention and emotional reactions to negative stimuli: The role of age and cognitive reappraisal. *Psychology and Aging* 32(6): 543.

Wu, Z., Ji, D., Yu, K., Zeng, X., Wu, D., Shidujaman, M. (2021). AI Creativity and the Human-AI Co-creation Model. In: Kurosu, M. (eds.) *Human-Computer Interaction. Theory, Methods and Tools. HCII 2021. Lecture Notes in Computer Science*, vol. 12762. Springer, Cham. **https://doi.org/10.1007/978-3-030-78462-1_13**

Xu, L., Becker, B., & Kendrick, K. M. (2019). Oxytocin facilitates social learning by promoting conformity to trusted individuals. *Frontiers in Neuroscience* 13: 56.

Yang, J., Li, Y., Calic, G., & Shevchenko, A. (2020). How multimedia shape crowdfunding outcomes: The overshadowing effect of images and videos on text in campaign information. *Journal of Business Research* 117: 6–18.

Yim, M. Y. C., Lee, J., & Jeong, H. (2021). Exploring the impact of the physical conditions of mannequin displays on mental simulation: An embodied cognition theory perspective. *Journal of Retailing and Consumer Services* 58: 102332.

Yuan, L., Kong, F., Luo, Y., Zeng, S., Lan, J., & You, X. (2019). Gender differences in large-scale and small-scale spatial ability: A systematic review based on behavioral and neuro-imaging research. *Frontiers in Behavioral Neuroscience* 13: 128.

Zajonc, R. B. (1968). Attitudinal effects of mere exposure. *Journal of Personality and Social Psychology* 9(2, Pt.2): 1–27. **https://doi.org/10.1037/h0025848**

Zak, P. J. (2015). Why inspiring stories make us react: The neuroscience of narrative. Cerebrum, 2015, cer-02-15.

Zaki, N. F., Spence, D. W., BaHammam, A. S., Pandi-Perumal, S. R., Cardinali, D. P., & Brown, G. M. (2018). Chronobiological theories of mood disorder. *European Archives of Psychiatry and Clinical Neuroscience* 268: 107–118.

Zatorre, R. J., Fields, R. D., & Johansen-Berg, H. (2012). Plasticity in gray and white: neuro-imaging changes in brain structure during learning. *Nature Neuroscience* 15(4): 528–536.

Zeeshan, M., & Obaid, M. H. (2013). Impact of Music on Consumer Behaviour: A Perspective on retail atmospheric. *Asian Journal of Business and Management Sciences* 3(2): 56–63.

Zhang, J., & Lee, E. J. (2022). "Two Rivers" brain map for social media marketing: Reward and information value drivers of SNS consumer engagement. *Journal of Business Research* 149: 494–505.

Zhang, K., Rigo, P., Su, X., Wang, M., Chen, Z., Esposito, G., . . . & Du, X. (2020). Brain responses to emotional infant faces in new mothers and nulliparous women. *Scientific Reports* 10(1): 9560.

Zhang, W., Liu, F., Zhou, L., Wang, W., Jiang, H., & Jiang, C. (2019). The effects of timbre on neural responses to musical emotion. *Music Perception: An Interdisciplinary Journal* 37(2): 134–146.

Zhao, Y., Sun, Q., Chen, G., & Yang, J. (2018). Hearing emotional sounds: Category representation in the human amygdala. *Social Neuroscience* 13(1): 117–128.

Żmudzka, E., Sałaciak, K., Sapa, J., & Pytka, K. (2018). Serotonin receptors in depression and anxiety: Insights from animal studies. *Life Sciences* 210: 106–124.

Zuo, H., Jones, M., Hope, T., & Jones, R. (2016). Sensory perception of material texture in consumer products. *The Design Journal* 19(3): 405–427.

Acknowledgments

Authoring a book like this – about the most significant technology of our lifetime, which is dramatically shaping the future of the twenty-first century for all humanity – demands a great deal of hard work.

And knowledge gleaned over decades. And painstaking research, pulling together the leading lights globally in neuroscience and GenAi.

And finally, a deep and abiding love for and fascination with the subject: the groundbreaking intersection of advanced neuroscience and GenAi.

So a book like this does not happen without the support, guidance, sponsorship, and devotion of a number of people. They are too numerous to list, but we single a few out for the enormity of their contributions.

Scholars who taught us: Professor Robert Knight at the University of California at Berkeley, Professor Rajiv Lal at the Harvard Business school, Professor Shankar Sastry at the University of California at Berkeley.

Colleagues who debated us: Freya, Satheesh, Sugumar, Ajit, Karthik, Sylvia, Sanjay, Ying, David, Anna, Eshaan, Tom, Bob, Lorinda, Beale, Srikanth, Daphne, Rob, Audrey, George.

Clients who shaped our thinking: Punit, Aparna, Khushboo, Priya, Christina, Vijay, Madhu, Fernando, Nitin, Matt, Nitesh, Katja, Ludo, Asmita, Myralda, Kavita, Isha, Kirti, Ram, Prat, Nari, Suman, Leena, Amrit, Aryan, David, Todd, Pankaj, Grace, Pavi, Jai, Sasha, Stan, Andrew, Ash, Krishnakumar, Russ, Nitin.

Accomplished leaders who took the time to talk to us and support us: Mukesh, Manoj, Carlos, Bill, AR, Jenny.

Friends in life who brought laughter to us: Caroline, Mara, Tristan, Ram, Wally, Jaweed, Rebecca, Raed, Kate, Jeff, Bruno, Darci, Maren, Nolie, Madeleine, Roland, Betty, Prasad, Sudha, Bhargavi, Gopal, Riya, Risha, Krithika, Sumanth, Sriman, Sarina, Habib, Abdul, Hussain, Jose, Andie.

Four-legged friends who brought us love: Osho, Teddy, Sky, and Zeus.

We acknowledge all of you for having taught us and supported us. Yes, dear reader, whatever you love in this book is because of the incredible teachings of all of these listed here and more. Whatever you find to be less than perfect in this book is entirely the flaw of us, the authors, and we humbly promise to do better the next time.

About the Authors

Dr. A. K. Pradeep

Dr. A. K. Pradeep is the founder and CEO of Sensori.Ai, and is one of the world's leading experts on applied neuroscience and artificial intelligence. He is a highly successful serial entrepreneur, having founded NeuroFocus (acquired by Nielsen) and BoardVantage (acquired by Nasdaq). Dr. Pradeep has over 90 patents and has written three books: *The Buying Brain, Ai for Marketing and Product Innovation,* and *Governance: 6 Easy Pieces.*

Dr. Pradeep was awarded the Grand Prize by the Advertising Research Foundation, and named Business Leader of the Year by the US India Business Council. He earned his PhD in engineering from the University of California at Berkeley with a specialization in nonlinear control systems.

Dr. Anirudh Acharya

Dr. Anirudh Acharya serves as the chief AI officer of Sensori.Ai, leading the development of all its innovation in artificial intelligence and tech. His journey started in academia focusing on mathematical physics, leading to research in the development of advanced techniques in quantum state detection and estimation. He has won several academic awards and grants for research excellence. He is also an accomplished visual artist, with successful gallery exhibitions and album artwork. He earned his PhD in mathematics from the University of Nottingham.

Dr. Rajat Chakravarty

Dr. Rajat Chakravarty is COO at Sensori.ai. With 10+ years technical and consulting expertise in data-driven analytics, he has led AI transformations with clients across healthcare, mortgage lending, advertising, and FMCG. He's been a radio broadcaster for four years, speaks six languages, and is often found either giving industry talks at universities in California or moshing at death metal concerts. He earned his PhD in engineering from the University of Saskatchewan in statistical analysis of turbulent flows, with two other degrees in aerospace engineering from IIT Bombay.

Ratnakar Dev

Ratnakar Dev is the CTO of Sensori.Ai. He has over 25 years of experience leading technical teams to successful acquisitions in the fields of collaboration, healthcare, AI, and neuro-marketing. He lives in the Silicon Valley Bay area with his wife and two children. He has a degree in computer engineering from the University of Pune.

Index

Page numbers followed by *f* refer to figures.